MW00760945

Springer Theses

Recognizing Outstanding Ph.D. Research

For further volumes:
http://www.springer.com/series/8790

Aims and Scope

The series "Springer Theses" brings together a selection of the very best Ph.D. theses from around the world and across the physical sciences. Nominated and endorsed by two recognized specialists, each published volume has been selected for its scientific excellence and the high impact of its contents for the pertinent field of research. For greater accessibility to non-specialists, the published versions include an extended introduction, as well as a foreword by the student's supervisor explaining the special relevance of the work for the field. As a whole, the series will provide a valuable resource both for newcomers to the research fields described, and for other scientists seeking detailed background information on special questions. Finally, it provides an accredited documentation of the valuable contributions made by today's younger generation of scientists.

Theses are accepted into the series by invited nomination only and must fulfill all of the following criteria

- They must be written in good English.
- The topic should fall within the confines of Chemistry, Physics, Earth Sciences and related interdisciplinary fields such as Materials, Nanoscience, Chemical Engineering, Complex Systems and Biophysics.
- The work reported in the thesis must represent a significant scientific advance.
- If the thesis includes previously published material, permission to reproduce this must be gained from the respective copyright holder.
- They must have been examined and passed during the 12 months prior to nomination.
- Each thesis should include a foreword by the supervisor outlining the significance of its content.
- The theses should have a clearly defined structure including an introduction accessible to scientists not expert in that particular field.

Julia Poncela Casasnovas

Evolutionary Games in Complex Topologies

Interplay Between Structure and Dynamics

Doctoral Thesis accepted by
the University of Zaragoza, Spain

Author
Dr. Julia Poncela Casasnovas
University of Zaragoza
Zaragoza
Spain

Supervisors
Prof. Luis Mario Floría
Department of Physics of Condensed Matter
University of Zaragoza
Zaragoza
Spain

Dr. Yamir Moreno Vega
Institute for Biocomputation and Physics of Complex Systems
University of Zaragoza
Zaragoza
Spain

Dr. Jesús Gómez-Gardeñes
Department of Physics of Condensed Matter
University of Zaragoza
Zaragoza
Spain

ISSN 2190-5053 ISSN 2190-5061 (electronic)
ISBN 978-3-642-30116-2 ISBN 978-3-642-30117-9 (eBook)
DOI 10.1007/978-3-642-30117-9
Springer Heidelberg New York Dordrecht London

Library of Congress Control Number: 2012938228

© Springer-Verlag Berlin Heidelberg 2012
This work is subject to copyright. All rights are reserved by the Publisher, whether the whole or part of the material is concerned, specifically the rights of translation, reprinting, reuse of illustrations, recitation, broadcasting, reproduction on microfilms or in any other physical way, and transmission or information storage and retrieval, electronic adaptation, computer software, or by similar or dissimilar methodology now known or hereafter developed. Exempted from this legal reservation are brief excerpts in connection with reviews or scholarly analysis or material supplied specifically for the purpose of being entered and executed on a computer system, for exclusive use by the purchaser of the work. Duplication of this publication or parts thereof is permitted only under the provisions of the Copyright Law of the Publisher's location, in its current version, and permission for use must always be obtained from Springer. Permissions for use may be obtained through RightsLink at the Copyright Clearance Center. Violations are liable to prosecution under the respective Copyright Law.
The use of general descriptive names, registered names, trademarks, service marks, etc. in this publication does not imply, even in the absence of a specific statement, that such names are exempt from the relevant protective laws and regulations and therefore free for general use.
While the advice and information in this book are believed to be true and accurate at the date of publication, neither the authors nor the editors nor the publisher can accept any legal responsibility for any errors or omissions that may be made. The publisher makes no warranty, express or implied, with respect to the material contained herein.

Printed on acid-free paper

Springer is part of Springer Science+Business Media (www.springer.com)

Als sie die Kommunisten geholt haben, hab ich nichts gesagt. Ich war ja kein Kommunist.
Als sie die Sozialdemokraten geholt haben,hab ich nichts gesagt. Ich war ja kein Sozialdemokrat.
Als sie die Juden geholt haben,hab ich nichts gesagt. Ich war ja kein Jude.
Als sie mich geholt haben, war niemand mehr da der hätte etwas sagen können.

Martin Niemöller (1892–1984).

A Olga, mi madre
Ojalá estuvieras aquí
♡

Supervisors' Foreword

Phenomena as diverse as the emergence of social consensus, the spreading of infectious diseases in animal and human populations, the adoption of social norms, or the transmission of cultural traits, can all be modeled as the dynamical interaction of macroscopically many entities or agents. Though referring to very different systems, the modeling of these and many other dynamical collective processes is a cross-disciplinary research field that has developed a growing set of methods and techniques of an impressive explanatory power, many of them inspired by, or akin to, the statistical and nonlinear physical analysis of equilibrium and non-equilibrium systems.

For a wide variety of issues and important questions in social, economical, and biological sciences a very successful formulation is that of the Evolutionary Game Theory, where interaction among agents is modeled as a game, with different possible strategies, from which agents obtain payoffs and strategies spread over the population (by e.g. learning, imitation, heredity, or other mechanisms) in direct proportion to the payoff obtained. The outcome of this feedback dynamics, i.e., the time evolution of the fraction of each type of strategists in the population depends in a crucial way (among other factors) on the social structure of interactions. This is the issue addressed in Part I of this thesis work, with the use of the Prisoner's Dilemma game, the archetype framework for the study of a long-standing issue in evolutionary theory, namely the evolution of Cooperation. In Part II the author takes an important step forward, by allowing the network of interactive contacts to evolve in time under the influence of the strategic field evolution, and so considering the co-evolution of strategies and social contacts. In doing so, this research contributes to one of the currently debated questions in network science, namely the origin of the commonly observed topological characteristics of real complex networks as, e.g., high degree heterogeneity, small-world property, clustering, and community structure.

Zaragoza, Spain, February 2012

Luis Mario Floría
Yamir Moreno
Jesús Gómez-Gardeñes

Contents

Chapter 1
Introduction

Many real systems from very different fields, such as food webs [1–3], the electrical power grids, the social entanglement of acquaintances [4], the Word Wide Web or the Internet [5–7], were almost intractable just a few years ago due to both their large number of individuals and the complexity of the pattern of connections among them. They all have been recently characterized as *networks* [8–13], opening a new and very promising subject for researchers all over the world.

In a few words, a network can be defined as a set of nodes or individuals, and a set of connections or links that represent some kind of physical or abstract relationship among them. Specifically, a network can be considered *complex* if it has a pattern of connections highly non trivial. These systems have found in Graph Theory a useful tool that allows us to study, analyze, reproduce and describe them accurately, extracting some common structural features to characterize and organize them accordingly. And surprisingly enough, most of real networked systems seem to share some of these structural features, regardless their particular origin, thus entitling this new discipline, far beyond simple anecdotal facts.

Other real examples [12, 14, 15] are neural networks of animals (where the nodes are neurons, and links represent chemical synapses), cellular and metabolic networks (where nodes stand for the different molecules or metabolites that take part on the system of chemical reactions, and a link between two of them means that one is the reactive and the other one is its product), the network of actors in Hollywood (two actors have a link if they have worked together in a film), the co-authorship and citation networks of scientists (similarly, two scientist will share a link if they have a common paper, or two papers will have a link between them if one cites the other, respectively), the air transportation network (nodes stand for airports and links represent direct flights between an origin and a destination) or the network of sexual human contacts (where a link binds two human beings that have had sex together).

On the other hand, the fact that all of them have complex structures has been proven to strongly affect the outcome of the great variety processes that may take place on top of them, in comparison with well-mixed situations or even lattice-like underlying structures. Thus, it modifies sometimes dramatically the assumptions as

J. Poncela Casasnovas, *Evolutionary Games in Complex Topologies*, Springer Theses,
DOI: 10.1007/978-3-642-30117-9_1, © Springer-Verlag Berlin Heidelberg 2012

well as the conclusions one can make from such systems. For example, the dynamics of disease spreading is very different depending on the social structure one considers for the propagation process (and so are the measures that should be taken in order to effectively fight it off), or when dealing with traffic jams in the road network or on the Internet, it is also essential to know the topology underneath, in order to design effective strategies.

In Fig. 1.1 we show some other examples of real networks: **(a)** represents the email network from the members of the Universitat Rovira i Virgili (Spain), where we can clearly see different branches (or communities), corresponding to different departments and areas within those departments [16], **(b)** is the network that combines local metropolitan commuters and long-range airline travelers during a global epidemic [17], and **(c)** shows the New Testament social network (http://www.esv.org/blog/2007/01/mapping-nt-social-networks/).

The first attempts to model such real networks were over-simplifying: lattices and regular random networks [18] were foremost used to try to encapsulate some of the basic characteristics of these complex networks. In a lattice, the individuals are arranged at regular distances in one, two or three spatial dimensions, with a fixed number of neighbors (or coordination number). On the other hand, random graphs are just a set of individuals with aleatory connections among them, but without any order or periodicity. One can characterize the distribution of probability for the number of those connections in the system by a Poisson distribution, so there is a well-defined mean value, or it can also be given by a Dirac-delta, which means that every element in the system has exactly the same number of neighbors. Nonetheless, the concept of dimensionality is hard to define in random graphs, and also in complex networks in general.

Obviously, and despite its undeniable importance as first attempts in the matter, these kind of models are unrealistic representations of real systems. Due to its lack of accuracy, they fail to explain some features such as the well-know small-world phenomenon or *six degrees of separation* [19, 20]. Roughly speaking, it implies that any two individuals in the network are likely to be connected through a very short sequence of intermediate acquaintances. This has been the subject of anecdotal observation and folklore for a long time: often we meet a stranger and discover, astonished, that we have an acquaintance in common. Nonetheless, it finally became a significant area of study in the social sciences, in large part through the striking experiments by Stanley Milgram in the 1960's [21]. Later on, it has been shown that many other real networked systems, such as technological or biological ones, display often this feature.

Besides, in these social networks, it is very likely that two different friends of a person have also met (it is to say, they have a high clustering coefficient). Moreover, these two properties usually appear simultaneously in real networks, so both should be taken into account if one wants to model reality with some accuracy. On the one hand, lattices achieve the second property, but not the first one, and for random topologies, it happens the other way around. Thus the next step was to try to model a network that combines both features, and the *Small-word* network [10] does it. This particular model was the first one to enclose simultaneously the two properties of

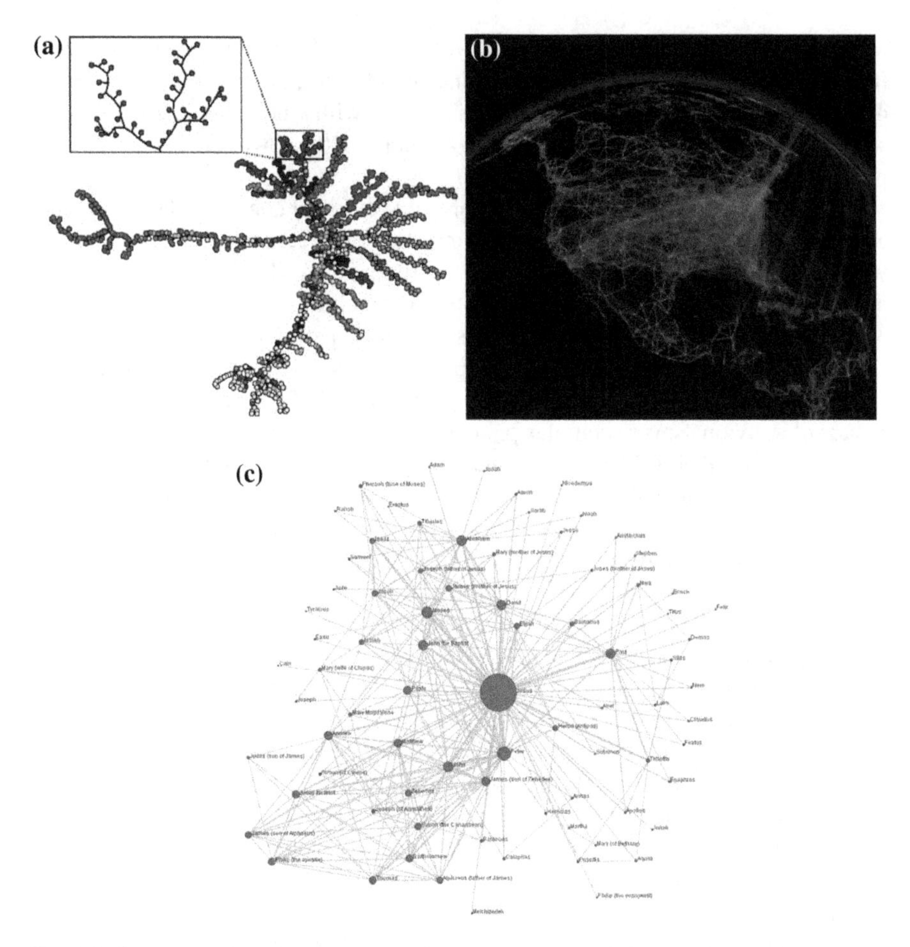

Fig. 1.1 Some examples of real networks: **(a)** the email network from the Universidad Rovira-Virgili (Spain) [16], **(b)** the network of local metropolitan commuters and long-range airline travelers during a global epidemic [17] and **(c)** the New Testament social network (obtained from the homepage of the English Standard Version Bible:http://www.esv.org/blog/2007/01/mapping-nt-social-networks/)

real networked systems mentioned before, and it is built as follows: starting from a regular lattice, and by randomly rewiring a certain percentage of the links, the network gets some shortcuts between otherwise distant nodes, so they will have a low value for the average path length, like random graphs, but still with a high value of the clustering coefficient, like lattices.

As an ulterior improvement in realism at modeling, one can consider yet another very common feature among real networks, that is the heterogeneity in the number of connections a node has: we all know people that are really popular, and some other people that are incurably unsociable. In the same way, there are a few very important airports and a lot of medium and even more small ones. None of the previous models accounted for this particular feature, and were the so-called *Scale-free* (SF) networks [8] the ones that did it. This particular kind of networks, have a power-law degree distribution (it is, the probability of finding a node with k neighbors), $P(k) \sim k^{-\gamma}$, with $2 < \gamma < 3$. Usually, real networks are not strictly power-lawed, but they do present some degree of heterogeneity.

As we have already mentioned, there are very different contexts where networks can appear (zoology, biochemistry, sociology, technology...) and so the processes that will take place on top of them can be very different as well [11, 12, 22, 23]: from disease [11, 12, 24–30] or rumor spreading to synchronization dynamics [12, 23, 31–34], traffic jams and cooperation. This last one is particularly interesting for us, since there are countless examples of cooperation in Nature: cells cooperate to form tissues, organs cooperate to form living organisms, and of course, when it comes to groups of individuals, very complex phenomena can arise: they can cooperate within a family to raise their offspring, form hunting parties, form alliances, stick together in order to reduce the risk of predation, and in general, to form societies...

However, why cooperation emerges and survives in hostile environments, when defecting is a much more profitable sort-term strategy, is a question that still remains open.

A lot of researchers are currently trying to answer that challenging question, and some key ideas have been pointed out so far, such as kin selection or the necessity to protect the offspring or the family in general (for obvious evolutionary reasons, or as the geneticist and evolutionary biologist J.B.S. Haldane said: 'I will jump into the river to save two brothers or eight cousins'). Also, there is a benefit in cooperating with someone you will probably meet again in the future (direct reciprocity) or if you gain some (good) reputation because of it (indirect reciprocity, see [35] and references therein). On the other hand, for repeated-encounter situations, where individuals have some kind of memory of the past or even plans for the future, there are some complex strategies that can be more successful than others...

Game theory attempts to mathematically capture the behavior of such individuals in strategic situations, in which their success in making choices (that is measured in terms of benefits) depends on the choices of others. Evolutionary Game Theory is a branch of Game Theory that studies the time evolution of large *populations* of individuals who repeatedly play a game and are exposed to evolutionary pressures (selection and replication, with or without mutation), and it has been proven to be the mathematical framework to deal with questions such as the problem of evolution of cooperation. Specifically, the Prisoner's Dilemma game has been widely used [35–45] as a perfect metaphor for the study of cooperation among individuals, where it is clearly more profitable to defect regardless the opponent's strategy, but also it would be better for the two adversaries if both of them decided to cooperate, instead of defecting.

We are interested in cooperation on very simple scenarios: when individuals have no memory or plan for the future at all, and they do not recognize their families nor have reputations to keep. Thus, we want to study the merely structural factors that can help cooperation in a given situation. It is already known that lattices promote cooperation [46–48], with respect to the all-to-all scenario, because it allow cooperators to for clusters and hence be more resiliant against defectors. Here we will focus more on the reasons why cooperation seems to be enhanced by heterogeneity in the number of connections [22, 49–59], compared to the more regular case of random topologies.

Therefore, in the first part of this Thesis, we will address the problem of the sustenance of cooperation in complex static topologies, comparing the dynamics on top of two fundamental kind of networks: random and scale-free. We will model the issue of choosing between cooperation and defection via the paradigmatic and well-known Prisoner's Dilemma game. This is a very simple 2×2 game where there are two players who can choose between two distinct strategies: cooperate and defect. And depending on its strategy and its opponent's choice, they will get an well-defined benefit (usually represented by a payoff matrix). Essentially, the problem is that, given the payoff matrix of this game, defecting is the safest strategy regardless the one the opponent chooses, but, if both decided to cooperate, they would get higher payoff than if both of them defect (hence, the dilemma).

Specifically, in the first part of this Thesis, we will use computer simulations to study how cooperators and defectors in the system, spontaneously and after a transient period of time, arrange themselves at a microscopic level, giving rise to very different organization patterns, which will be at the root of the very distinct levels of average cooperation found in different kinds of networks.

On the other hand, we are well aware that real networks are not static entities at all: not only there can be different dynamics evolving on top of them, but also the structure of the network itself usually changes over time. New nodes can enter the system, others can disappear and also new connections can be established or erased. Moreover, the processes that take place on top of them can shape the topology, and the other way around as well. So, we consider that a natural next step in our study of cooperation in complex networks should be a model where the dynamics and the growth of the network are entangled. In this way, the second part of this Thesis will be devoted to developing two different models of growing networks that reflect some of the characteristics of an evolving real network. Thus, in both our models, the outcome of the dynamics will be taken into account for the growing process. Specifically, the dynamics will be again the Prisoner's Dilemma game, and the payoff obtained by the nodes will affect its capability of attracting links from the newcomers. Nonetheless, the two models differ in the kind of dependence between the probability of attachment of the new nodes with the payoff of those already present in the system, and also in the way a node evaluates whether to keep its current strategy or not by comparing with its neighbors will also be different in both models. Besides, we will analyze, along with the average levels of cooperation achieved in every case, the structures that can emerge from these combined processes, depending on the specific values of the parameters of the model. In order to do that, we will measure the relevant topological magnitudes, such as the degree distribution, the average path length and the clustering coefficient of the resulting networks. Moreover, we will establish some comparisons between the results obtained with these models, when the final size is achieved, and those known for fixed-size static networks, such as Erdös-Rényi (ER) random networks, Barabási-Albert (BA) scale-free networks and random scale-free networks. In summary, with this work we hope contribute to solve the open question of how cooperation is affected by the underlaying topological structure of the population.

References

1. J. Dunne, R. Williams, and N. Martinez., Marine Ecological Press Series **273**, 291 (2004).
2. J. Cohen, F. Briand, and C. Newman, *Community food webs: data and theory.* (Springer-Verlag, New York, 1990).
3. R. Williams and N. Martínez, Nature **404**, 108 (2000).
4. B. Skyrms and R. Pemantle, Proc. Natl. Acad. Sci. USA **97**, 9340 (2000).
5. R. Pastor-Satorras and A. Vespignani., *Evolution and Structure of the Internet: A Statistical Physics Approach.* (Cambridge University Press, Cambridge, 2004).
6. A. Brodera, R. Kumarb, F. Maghoula, P. Raghavanb, S. Rajagopalanb, R. Statac, A. Tomkinsb, and J. Wienerc, Graph structure in the Web. Comput. Newt. **33**, 309 (2000).
7. M. Faloutsos, P. Faloutsos, and C. Faloutsos, Comput. Commun. Rev. **29**, 251 (1999).
8. A. L. Barabási and R. Albert, Science **286**, 509 (1999).
9. S. H. Strogatz, Nature **410**, 268 (2001).
10. D. J. Watts and S. H. Strogatz, Nature **393**, 440 (1998).
11. M. Newman, SIAM Review **45**, 167 (2003).
12. S. Boccaletti, V. Latora, Y. Moreno, M. Chavez, and D. U. Hwang, Phys. Rep. **424**, 175 (2006).
13. R. Albert and A. L. Barabási, Rev. Mod. Phys. **74**, 47 (2002).
14. T. B. Achacoso and W. S. Yamamoto, *AY's neuroanatomy of C. elegans for computation.* (CRC Press, Boca Raton, Fl, 1992).
15. H. Jeong, B. Tombor, R. Albert, Z. N. Oltvai, and A.-L. Barabási, The large-scale organization of metabolic networks. Nature **407**, 651 (2000).
16. A. Arenas, L. Danon, A. Díaz-Guilera, P. Gleiser, and R. Guimerá. Community analysis in social networks. European Physical Journal B **38**(**2**), 373 (2004).
17. D. Balcan, V. Colizza, B. Goncalves, H. Hu, J. J. Ramasco, and A. Vespignani, Proc. Natl. Acad. Sci. USA 106, 21484 (2009).
18. P. Erdős and A. Renyi, Publicationes Mathematicae Debrecen, **6**, 290 (1959).
19. S. Milgram, Psycol. Today, **2**, 60 (1967).
20. J. Guare, *Six degrees of separation: a play.* (Vintage Books, New York, 1990).
21. J. Travers and S. Milgram, Sociometry, **32**, 425 (1969).
22. G. Szabó and G. Fáth, Evolutionary games on graphs. Phys. Rep. **446**, 97 (2007).
23. A. Arenas, A. Díaz-Guilera, J. Kurths, Y. Moreno, and C. Zhou, Phys. Rep. **469**, 93 (2008).
24. R. Pastor-Satorras and A. Vespignani, Phys. Rev. Lett. **86**, 3200 (2001).
25. R. Pastor-Satorras and A. Vespignani, Phys. Rev. E, **63**, 066117 (2001).
26. Y. Moreno, R. Pastor-Satorras, and A. Vespignani, European Physical Journal B, **26**, 521 (2002).
27. R. Pastor-Satorras and A. Vespignani, Phys. Rev. E, **65**, 036104 (2002).
28. M. Boguñá, R. Pastor-Satorras, and A. Vespignani, Phys. Rev. Lett. **90**, 028701 (2003).
29. R.M. May and A.L. Lloyd, Phys. Rev. E **64**, 066112 (2001).
30. M. Newman, Phys. Rev. E **66**, 016128 (2002).
31. J. Gómez-Gardeñes, Y. Moreno, and A. Arenas, Phys. Rev. Lett. **98**, 034101 (2007).
32. L. Donetti, P. I. Hurtado, and M. A. Muñoz, Phys. Rev. Lett. **95**, 188701, (2005).
33. J. Gómez-Gardeñes and Y. Moreno, Int. J. Bifurcation Chaos, **17**, 2501 (2007).
34. J. Gómez-Gardeñes, Y. Moreno, and A. Arenas, Phys. Rev. E **75**, 066106 (2007).
35. M. A. Nowak and K. Sigmund, Nature 437, 1291, 2005.
36. R. Axelrod, *The Evolution of Cooperation.* (Basic Books, New York, 1984).
37. W. Hamilton, J. Theor. Biol. **7**, 1 (1964).

38. R. Axelrod and W. D. Hamilton, Science **211**, 1390 (1981).
39. M. Nowak, Science **314**, 1560 (2006).
40. J. Hofbauer and K. Sigmund, *Evolutionary games and population dynamics.* (Cambridge University Press, Cambridge, UK, 1998).
41. J. Hofbauer and K. Sigmund, Bull. Am. Math. Soc. **40**, 479 (2003).
42. M.A. Nowak and K. Sigmund, Nature **355**, 250 (1992).
43. M.A. Nowak and K. Sigmund, Acta Applicandae Math, **20**, 247 (1990).
44. R. Axelrod, *The complexity of cooperation: agent-based models of competition and collaboration.* (Princeton University Press., Princeton, NJ, 1997).
45. M.A. Nowak, *Evolutionary dynamics: exploring the equations of life.* (Harvard University Press., Cambridge, MA, 2006).
46. M. A. Nowak and R. M. May, Nature **359**, 826 (1992).
47. M. Nowak, S. Bonhoeffer, and R. May, Int. J. Bifurcation Chaos, **4**, 33 (1994).
48. M.A. Nowak, S. Bonhoeffer, and R.M. May, Proc. Natl. Acad. Sci. USA **91**, 4877 (1994).
49. F. C. Santos and J. M. Pacheco, Phys. Rev. Lett. **95**, 098104 (2005).
50. F. C. Santos, F. J. Rodrigues, and J. M. Pacheco, Proc. Biol. Sci. **273**, 51 (2006).
51. F. C. Santos and J. M. Pacheco, J. Evol. Biol. **19**, 726 (2006).
52. F. C. Santos, J. M. Pacheco, and T. Lenaerts, Proc. Natl. Acad. Sci. USA **103**, 3490 (2006).
53. H. Ohtsuki, E. Lieberman C. Hauert, and M. A. Nowak, Nature **441** 502 (2006).
54. G. Abramson and M. Kuperman, Phys. Rev. E **63**, 030901(R) (2001).
55. V. M. Eguí
56. T. Killingback and M. Doebeli, Proc. R. Soc. Lond. **263**, 1135 (1996).
57. A. Szolnoki, M. Perc, and Z. Danku, Physica A, **387**, 2075 (2008).
58. J. Vukov and G. Szab6and A. Szolnoki, Phys. Rev. E **77**, 026109 (2008).
59. J. G6mez-Gardeñes, M. Campillo, L. M. Floría, and Y. Moreno, Phys. Rev. Lett. **98**, 108103 (2007).

Chapter 2
Some Basic Concepts on Complex Networks and Games

Since this thesis is mainly devoted to the study of one particular game, the Prisoner's Dilemma, on complex networks (static ones in the first part of it, and two more sophisticated models that combine the growth with the play in the second), we consider that it is useful to state and explain first some notions on both networks and games. So, in this chapter, we want to provide just a few very basic concepts and definitions on Complex Networks and Game Theory that we will use later on during the full elaboration of this thesis. We hope they will help setting the foundations to understand our work perfectly, so the reader will not need any external help to comprehend, and also it will serve as an introduction to the two fundamental components on which this thesis is based.

2.1 Complex Networks

The study of complex networks is a relatively recent field, and it has been inspired by the observation of many real systems, such as biological, social or technological ones. In the first part of this chapter we want to give a few examples of real networks, just to motivate the study of such structures, by establishing its ubiquity in natural and artificial systems. Then, we will give some of the basic definitions needed in order to properly describe networks [1], such as the degree of a node, the degree distribution of a network, the clustering coefficient or the average path length. Then, we will explain some useful models for building different kinds of graphs, such as the Erdös and Rényi (ER), the Barabási-Albert (BA) or the Small-World by Watts and Strogatz model. Finally, we will mention some of the many possible processes that can take place on top of complex networks.

J. Poncela Casasnovas, *Evolutionary Games in Complex Topologies*, Springer Theses,
DOI: 10.1007/978-3-642-30117-9_2, © Springer-Verlag Berlin Heidelberg 2012

2.1.1 Examples of Real Networks

As it has been pointed out along the Introduction of this thesis, many real systems [1, 2] can be described as complex networks, and this relatively new approach can provide new insights to better understanding, and tools to deal with unsolved problems. In very different fields, such as biology, immunology, sociology, technology or economics, there are plenty of examples of networks. In every particular field, both the nodes and the links of the networks will represent completely different things, but the fact that this kind of structures are so ubiquitous in Nature, is surprising and very promising.

One can consider technological structures, such as the air transportation networks for a particular region or for the whole planet, where the nodes are airports and the links represent direct flights between them, the road networks connecting cities or the power grids that supply electricity to a country, with its power stations represented by nodes and the links standing for the wires. There is also the WWW, where nodes are web pages connected by hyperlinks, or the Internet (see Fig. 2.1 (**Left**)), made up of billions of hosts, physically connected among them. Since modern societies depend strongly on these infrastructures, it is obviously very important to have detailed information about them, in order to be able to predict its behavior or act correctly during a crisis.

In biology, there are several examples as well, like food webs on an ecosystem (see Fig. 2.1 (**Right**)), or on a more basic level, the metabolic networks of different processes. On the other hand, maybe some of the more tangled complex networks one can consider (from the point of view of both number of interconnections and variability over time) are those that describe social relationships, where nodes are people, and links represent some kind of interaction: from groups of mere friends, people with similar interests or collaborators in some particular field [4, 5] (scientific collaborations or citations, or networks of musicians that play together regularly,...), to sexual contact networks or new global phenomena like Facebook, MySpace or Twitter. It is probablyl because of the complex nature of the human being itself, that such social structures are often so entangled and fascinating.

On the other hand, we want to point out that, when dealing with real networks one has to take into account that the available data can (and probably will) have mistakes: there can be missing or spurious nodes or links. Some effort has been put to try to obtain the 'real network' and its topological properties out of the observational data (see for example [6]).

Finally, the kind of processes that will take place on top of them can be very diverse (synchronization, traffic of information or of something else, disease or rumor spreading, games, learning processes...), but it is very useful to be able to characterize them structurally as precisely as possible first, trying to find out what are the main and more relevant features all of them share, if any. Moreover, as we will see later on, the structure will be a key factor in the outcome of any dynamical process that will take place on top of such structured systems. Thus, we will address next the topological characterization of complex networks.

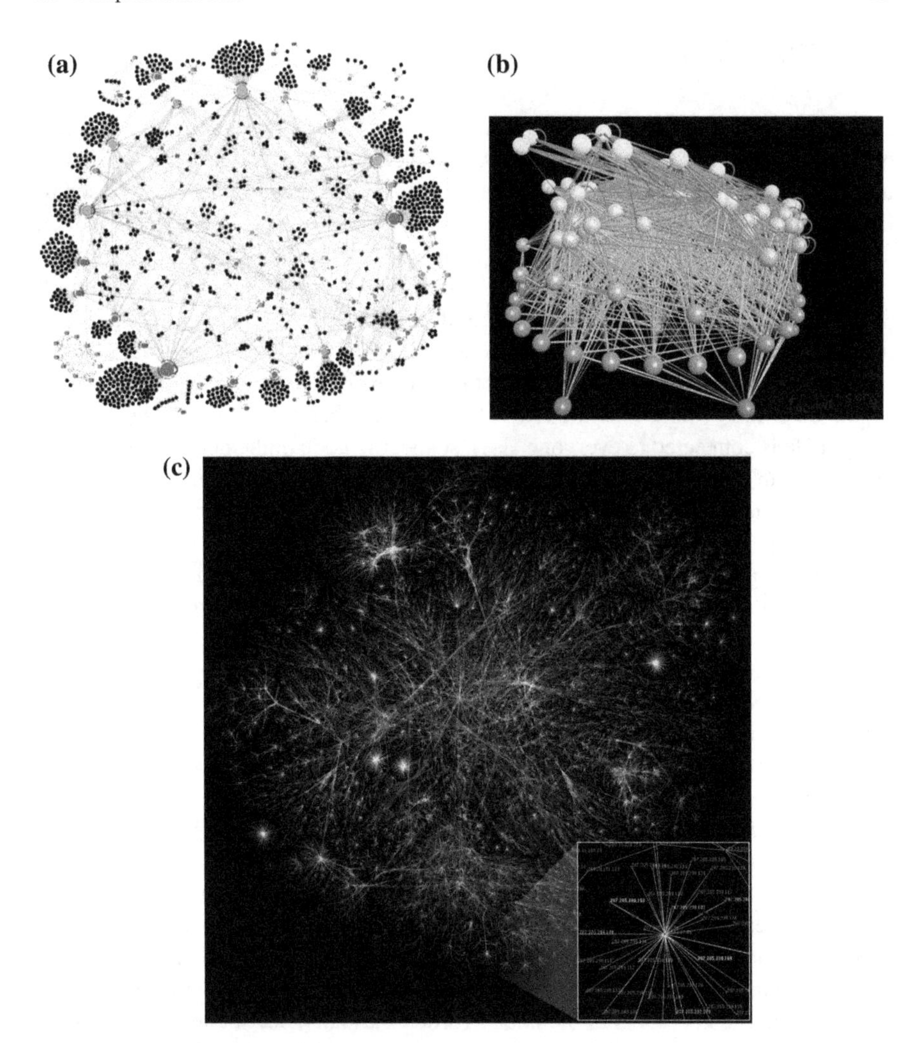

Fig. 2.1 **a** Gene regulation network for the Mycobacterium Tuberulosis. Every node represent a gene, and the links stand for the regulation relationship between a transcription factor and the correspondent regulated gene. Different colors mean different character of the genes, as far as regulation dynamics is concern [2]. **b** Food web of the Caribbean coral reef located in the Puerto Rico Virgin Islands. Node color represents trophic level: *red nodes* represent basal species, such as plants and detritus, *orange nodes* represent intermediate species, and *yellow nodes* represent top species or primary predators. Links characterize the interaction between two nodes, and the width of the link attenuates down the trophic cascade, so a link is thicker at the predator end and thinner at the prey end (Original image from [3], and generated by FoodWeb3D). **c** Visualization of a portion of the Internet, using over $5 \cdot 10^6$ edges. The colors represent different geographical regions. In the inset it is shown a particular node and its neighborhood. (Original image from 'The Opte Project': http://www.opte.org)

2.1.2 Definitions

A network is a set of items (called *nodes*, points or vertexes), with some connections between them (*links*, lines or edges). A complex network is a network with non-trivial topological features, i.e. its structure is irregular and complex as opposed to lattices, for example, that present total spatial regularity, or they can even evolve, adding and/or losing nodes and/or links over time.

Mathematically, we can represent a network using graph theory. A graph $\mathcal{G} = (\mathcal{N}, \mathcal{L})$, consists of two sets, $\mathcal{N}$ and $\mathcal{L}$, where $\mathcal{N} = \{n_1, n_2, \ldots, n_N\}$ are the nodes, and $\mathcal{L} = \{l_1, l_2, \ldots, l_K\}$ are the links. Obviously, N is the total number of nodes of the network, and K is the total number of links, which has to be a non-negative number, whose maximum is $N(N-1)/2$ (when the graph is *complete*, i.e. every node is connected to everyone else). A specific node of the network is denoted by a label i in the set $\mathcal{N}$. On the other hand, every link connects a pair of elements of $\mathcal{N}$, i and j, and is denoted by l_{ij}. Thereby, the pair of nodes i and j are called adjacent or neighbors. The usual way of representing a network graphically is by drawing a dot for every node and a line for every link that connects a pair of nodes. In addition to this, we can also define a subgraph $\mathcal{G}' = (\mathcal{N}', \mathcal{L}')$, of the graph $\mathcal{G} = (\mathcal{N}, \mathcal{L})$, if $\mathcal{N}' \subseteq \mathcal{N}$ and $\mathcal{L}' \subseteq \mathcal{L}$. A special case would be the subgraph of all the neighbors of a given node i and its corresponding links, denoted by $\mathcal{G}_i$. On the other hand, a graph is said to be *connected* if, for every pair of nodes i and j, there is a path to go from one to the other. If there is not such a path for at least one pair of nodes, then the graph will be *disconnected* or unconnected, and it will have therefore, two or more disconnected subgraphs.

Besides, another very useful way of representing a network is by using matrix representation. Given a graph $\mathcal{G} = (\mathcal{N}, \mathcal{L})$, the adjacency matrix $\mathcal{A}_{ij}$ is a $N \times N$ square matrix, whose entry a_{ij} $(i, j = 1, 2, \ldots, N)$ is equal to 1 when the link l_{ij} exists, and zero otherwise. Nonetheless, for implementation or practical purposes, we can use the connectivity matrix $\mathcal{C}_{ij}$ of the graph, that is a $Nxk_{\max}$ matrix, where $k_{\max}$ is the maximum connectivity of the nodes of the graph, and where the row i of it contains all the neighbors of the node i (ordered usually, but not necessarily, from the first to the last to connect with it when constructing the network). And we can also define a matrix of the pairs of neighbors, $\mathcal{D}_{ij}$, which is a $Lx2$ matrix, whose entries d_{l1} and d_{l2} are the pairs of nodes that are neighbors, with $l = 1, 2, \ldots, L$, and being L the total number of links in the network. The definition of these two matrices is not for rigorous mathematical purposes, but nonetheless, they will be very useful in order to implement them on programs and numerical simulations.

Degree of a Node and Degree Distribution of a Network

The *degree* or *connectivity* of a node is the number of neighbors it has. Using the adjacency matrix, we can formally define the degree of a node as:

$$k_i = \sum_{j \in \mathcal{N}} a_{ij} \tag{2.1}$$

If the graph is *directed*, then k_i will have two components: the ingoing links $k_i^{in} = \sum_j a_{ij}$ and the outgoing links $k_i^{out} = \sum_j a_{ji}$, so the total degree will be $k_i = k_i^{in} + k_i^{out}$.

On the other hand, the most basic topological characterization of the network as a whole is the *degree distribution*. We can define the degree distribution of the graph, $P(k)$, as the fraction of nodes in the network that have connectivity k, or equivalently, the probability that a node randomly chosen from the network has k neighbors. For example, random graphs (also known as 'one-peaked' or 'single-scaled') have a Poissonian degree distribution, while the $P(k)$ for a so-called scale-free network is a power law. For directed graphs, we will have two different distributions, $P(k^{in})$ and $P(k^{out})$.

The mean degree of a graph, $\langle k \rangle$ is the first moment of the degree distribution:

$$\langle k \rangle = \sum_k k P(k) \tag{2.2}$$

Furthermore, the second moment of the distribution, $\langle k^2 \rangle$ is the measure of the fluctuations of the degree distribution. As we will see later on, $\langle k^2 \rangle$ diverges in the limit of infinite graph size for scale-free graphs for certain values of the exponent of the power-law distribution, which is a very interesting property, that affects greatly the outcome of the dynamics that can take place on top of such topologies. For an *uncorrelated graph*, i.e. if the degree of every node is completely independent of its neighbors', then the degree distribution $P(k)$ is enough to describe the statistical properties of the network. But if the network is *correlated*, as it usually happens in many real systems, then the probability that a node of degree k has a neighbor with connectivity k', depends on k. In that case, we can define the conditional probability $P(k'|k)$, that a node with connectivity k has a neighbor with connectivity k'. We can also calculate the average degree of the nearest neighbor of nodes with degree k, given by:

$$k_{nn}(k) = \sum_{k'} k' P(k'|k) \tag{2.3}$$

So when the network is uncorrelated, obviously, we have that $k_{nn}(k)$ is independent of k, and equal to $k_{nn}(k) = \langle k^2 \rangle / \langle k \rangle$, but when it is correlated, then we can have *assortative* networks, if $k_{nn}(k)$ is an increasing function of k, or *disassortative* ones, when $k_{nn}(k)$ is a decreasing function of k. The first case implies that nodes tend to be linked with others with similar connectivity, whereas in the second one, the highly connected ones are mostly linked to the poorly connected ones.

Weighted and Directed Networks

Depending on the kind of interaction a link describes within the network, it can be weighted or non-weighted, directed or non-directed, and so will be the network.

If all the interactions in the network are alike, or in other words, when a link only establishes the presence of an interaction between two nodes, then the network is *non-weighted*. Otherwise, if there are different types of interactions, for example, some more important, or more frequent than others, then the links are *weighted*, and so is the graph. In this case, in addition to give the set of nodes and links of the network, we need to specify also the weight of every link in order to properly define a graph. So now we have: $\mathcal{G} = (\mathcal{N}, \mathcal{L}, \mathcal{W})$, where $\mathcal{W} = \{w_1, w_2, \ldots, w_K\}$ is the set of weights, that are real numbers attached to the corresponding links. Usually, they will be positive numbers, so the higher the value, the stronger the link between the pair of nodes, but also negative links have been used, describing some kind of repulsive interaction, for example [7]. On the other hand, if a link l_{ij} represents that i interacts with j and vice versa, then it is called *undirected*, but if in a system i can interact with j without j interacting necessarily with i, then in order to describe it correctly, we need *directed links*. In this case, the adjacency matrix will not be symmetric, in general.

Average Path Length, Betweenness and Clustering Coefficient

Given a particular network, it would be interesting to know the minimum distance between every pair of nodes, i.e. the shortest path lengths or geodesics. The knowledge of this information concerning a network can be useful for some processes that could take place on top on it (such as information traffic on the Internet, or rumor spreading on a social club), in order to work the best they can. Thus, we can define a square matrix $\mathcal{D}$, of size $N \times N$, whose entry d_{ij} is the minimum distance between the nodes i and j. On the one hand, the maximum of these d_{ij} is called the *diameter* of the graph, but a more useful magnitude to characterize the network, is the *average path length*, defined as the mean value of the geodesics between every pair of nodes in the network:

$$L = \frac{1}{N(N-1)} \sum_{i,j \in \mathcal{N}, i \neq j} d_{ij} \tag{2.4}$$

One can also ask how important or 'central' a particular node is within a graph, meaning how many shortest paths, or geodesics go through it. Thus, we can give a measure of the centrality of a node, by defining its *betweenness*:

$$b_i = \sum_{j,k \in \mathcal{N}, j \neq k} \frac{n_{jk}(i)}{n_{ij}}, \tag{2.5}$$

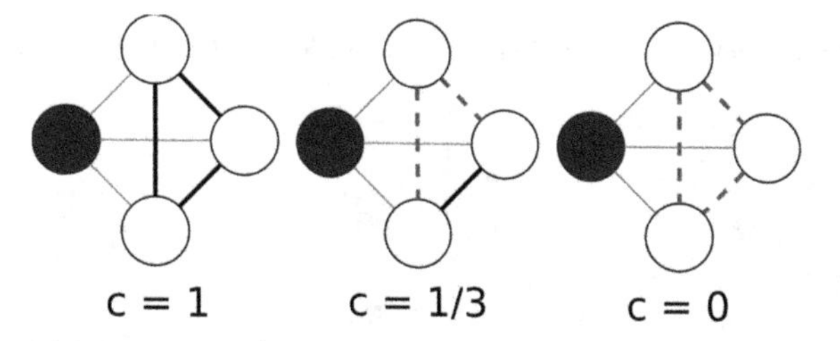

Fig. 2.2 Examples of the local clustering coefficient (for the *dark node*) for different connecting situations. It is computed as the proportion of connections among its neighbors which are actually realized (*thick black lines*) and the number of all possible connections, which in this particular example, is three. For every situation, the missing links are represented with dashed lines

where n_{jk} is the total number of geodesics connecting the nodes j and k, and $n_{jk}(i)$ is the number of geodesics connecting the nodes j and k that go through the node i.

The betweenness is a useful magnitude when constructing community detection algorithms [8, 9].

Clustering, or transitivity of a node, is a measure of how many triangles are there on the graph, or in other words, how likely is that, if a node i has two neighbors, say j and k, then the nodes j and k are also linked to each other. First, given a node i and the subgraph of its k_i neighbors, G_i, we can define the local clustering coefficient of node i as the ratio between the actual number of edges in the subgraph, e_i, and the maximum possible number of them in G_i:

$$c_i = \frac{2e_i}{k_i(k_i - 1)} = \frac{\sum_{j,m} a_{ij} a_{jm} a_{mi}}{k_i(k_i - 1)} \tag{2.6}$$

where a_{ij} are the entries of the adjacency matrix, defined at the beginning of this section. On Fig. 2.2 we show a diagram of how to calculate it for three very simple cases.

And then, we can define the clustering coefficient of the whole network, as the average of c_i over all the nodes in it:

$$C = \frac{1}{N} \sum_{j \in \mathcal{N}} c_i \tag{2.7}$$

Notice that, by definition, both the local and the global clustering coefficient satisfy: $0 \leq c_i \leq 1$ and $0 \leq C \leq 1$. As we will see, SF networks have low values for the average path length, but relatively high values for the clustering coefficient, while random topologies have low values for both magnitudes.

Finally, is worth mentioning that a power-law dependence of the clustering coefficient with the degree of the node ($C \sim k^{-1}$) is typical of a hierarchical organization on the network, which implies that sparsely connected nodes are part of highly clustered

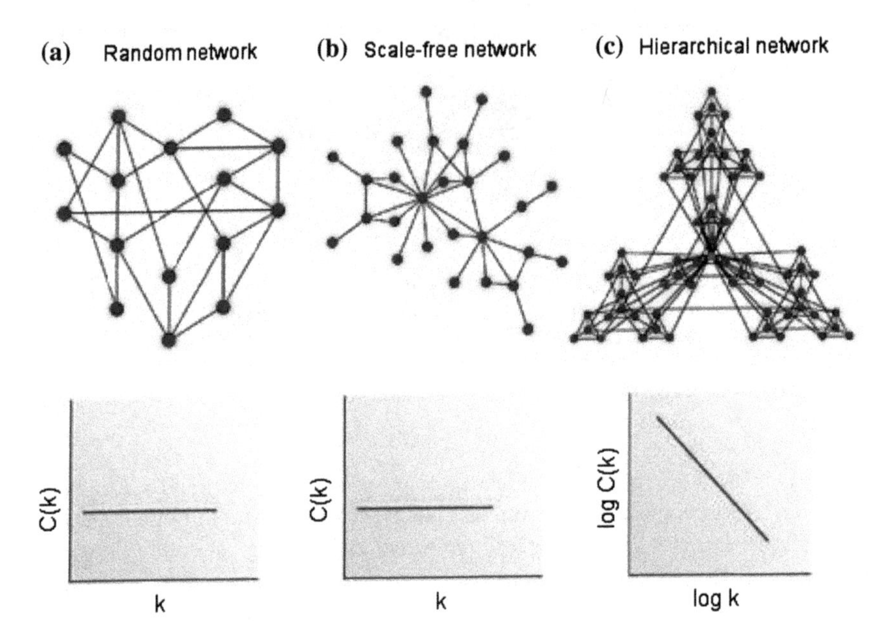

Fig. 2.3 Diagram with some examples of networks, specifically random (**a**), scale-free (**b**) and hierarchical ones (**c**), and its corresponding plots of the clustering coefficient versus the degree of the nodes. This dependence is a power-law for the hierarchical structures, while for the other two types, it is clearly independent. Original figure from [10]

areas, with communication between these different highly clustered neighborhoods being maintained by a few hubs (see Fig. 2.3).

Motifs and Communities on Networks

A *motif* is a n-noded pattern of connections (a subgraph) in a network that appears at a much higher rate than expected in a randomized version of the same network (see Sect. 5.1 for a detailed explanation of the randomizing procedure). Some real networks, such as the metabolic ones, display characteristic motifs, that seem to be specific of each kind of network. On Fig. 2.4 we show as an example, all the possible motifs for a 3-noded directed subgraph. Note that the number of n-noded motifs increases rapidly with n.

On the other hand, we can define a *community* within a network $\mathcal{G} = (\mathcal{N}, \mathcal{L})$, as a subgraph $\mathcal{G}' = (\mathcal{N}', \mathcal{L}')$ or a set of nodes, that are much more connected among themselves than with the rest of the network. Using just the sense that the intra-community connections are denser than the inter-community ones is of course a qualitative way of describing it. Nonetheless, to be able to detect such structures efficiently, a magnitude has been introduced to determine whether of not a partition of a network into communities is accurate enough: the *modularity*.

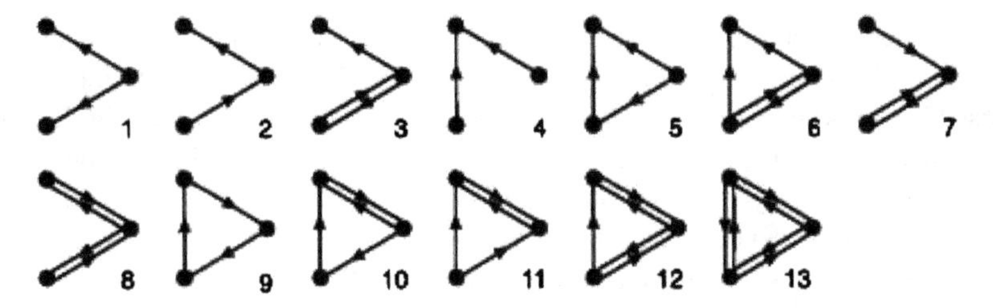

Fig. 2.4 All the possible 3-noded motifs on a directed network

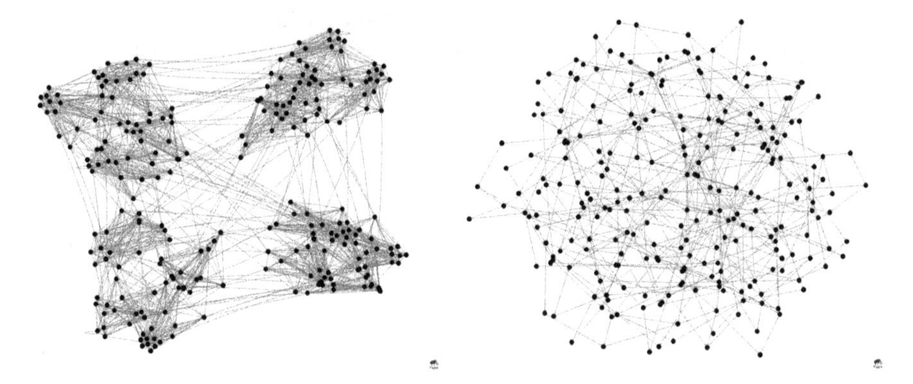

Fig. 2.5 Some examples of a network with (*left*) and without (*right*) community structure, both with $N = 256$ nodes. Original data of the community network created by Dr. L. Izquierdo (http://luis.izqui.org/communities/redes.zip)

Given an arbitrary network, and an arbitrary partition of it into N_c 'communities' (and this time, by this term we mean artificial communities, just a way to part the graph), we can build a $N_c \times N_c$ matrix whose entries e_{ij} are the ratio between the number of links starting at a node in community i and ending at a node in community j, and the total number of links present on the network (so the sum of any row or column, $a_i = \sum_j e_{ij}$, is the fraction of links connected to the community i).

In the case of a random partition of the network i.e., if it does not correspond to the actual community structure, or also if the network itself does not have a community structure (see Fig. 2.5 for some examples of networks with and without community structure), then the fraction of links within communities can be estimated as the probability that a link begins at a node in partition i, a_i, multiplied by the fraction of links that end at a node in partition i, also a_i, so the expected number of intra-community links is just $a_i a_i$. We also know the actual fraction of links exclusively within a partition, e_{ii}, so now we can compare the two values, and thus, we can define the modularity for a specific partition of our network as [8]:

$$Q = \sum_{i}^{N_c} (e_{ii} - a_{ij}^2) \tag{2.8}$$

Obviously, the closer to 1 the value of the modularity is, the more accurate the partition we have made of the network into communities. It is worth noticing that it is possible to find partitions of random networks that display relatively high values of modularity (up to $Q \sim 0.2$). The reason for this is that random graphs might have some community structure, just due to fluctuations. Moreover, it is important to stress that the presence of communities on a network can not be detected just via its degree distribution, so we can have two graphs with the same $P(k)$, one of them with community structure, and the other one without it.

One can easily realize that the space of possible partitions of a given network into communities is huge, so in order to effectively explore the landscape of values of Q, and find an accurately enough partition, we will need the help of some optimization techniques. For some very nice works on different community detection algorithms, see [8, 9, 11, 12] and references therein.

Finally, we want to mention that it is also possible to consider complex topologies with hierarchical structure, it is to say, networks that have communities within the communities. In this situation, we deal with several levels of description for the structure of the system (multiscale representation) [**?**]. Also, one can have a system with communities, where there is some degree of overlapping among them, and this fact will make it harder to detect accurately [13].

2.1.3 Some Network Models

In this section we want to present just a few models for growing networks. Specifically, we will address the models to build two of the most used kinds of networks: the ER and the BA model for random and scale-free networks respectively, since we will use them often, later on in this thesis, and also the well-known Small-World model by Watts and Strogatz. On the other hand, we will explain the Gardeñes-Moreno (GM) model, which interpolates between the ER and the BA model, because we will use it also in some chapters to come.

The ER Model

Erdös and Rényi proposed a model (ER) [14] to generate random graphs with N nodes and K links, where the term *random* refers to the disordered nature of the arrangement of links between different nodes. There are two possible ways of constructing such networks: in the first one, we start with N disconnected nodes and choose K pairs randomly, to link them with a probability $0 < p < 1$, avoiding multiple connections between two nodes, and also self-links. The alternative procedure is to start with N disconnected nodes, and link every possible couple with probability $0 < p < 1$. While the first option gets different networks with exactly K links and an average degree of $\langle k \rangle = 2K/N$, the second, gets networks with different number of connections, an average degree $\langle k \rangle = p(N-1)$, and the probability of having exactly

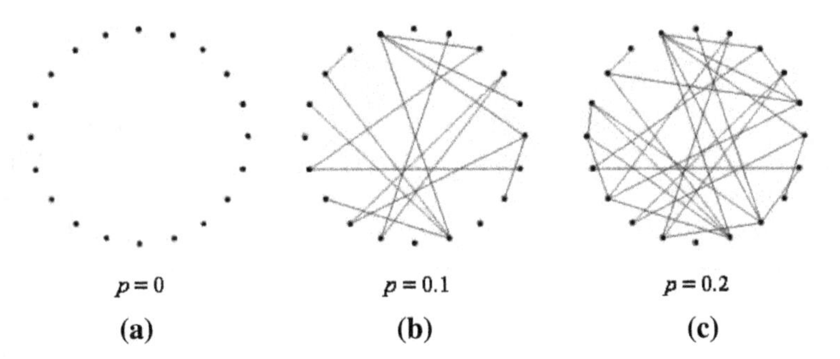

Fig. 2.6 Diagram of the ER model for random networks with $N = 20$ nodes

K links in a particular realization of the network is $p^K(1-p)^{N(N-1)/(2-K)}$. Nonetheless, both models coincide in the limit of large N, or thermodynamic limit. The probability of finding a node with a large connectivity decreases exponentially with K, so vertexes with large connectivity, $K \gg 1$, are practically absent (Fig. 2.6).

If one starts increasing the value of the probability of connection, from $p = 0$ (nodes totally disconnected) to $p = 1$ (complete graph), there is an interesting change of behavior at the critical value $p_c = 1/N$. Thus, if $p < p_c$, the graph is not connected (it has no component of size greater than $\mathcal{O}(lnN)$), if $p > p_c$, then the graph has a component of $\mathcal{O}(N)$, and the transition at p_c displays the typical features of a second phase transition. On the other hand, the probability of having a node with $k = k_i$ connections follows the Binomial distribution:

$$P(k = k_i) = C_{N-1}^{k} p^k (1-p)^{N-1-k} \tag{2.9}$$

where p^k is the probability of having k edges, $(1-p)^{N-1-k}$ is the probability of the absence of the remaining $(N-k)$ links, and C_{N-1}^{k} is the number of different ways of selecting the end points of these k nodes. Notice that, since all nodes of the networks are equivalent, this probability $P(k = k_i)$ is also the probability of choosing randomly a node with k_i neighbors. In the limit of large N and fixed $\langle k \rangle$, the degree distribution of the network can be accurately described by the Poisson distribution:

$$P(k) = e^{-\langle k \rangle} \frac{\langle k \rangle^k}{k!} \tag{2.10}$$

Moreover, for this particular topology, the dependence of the clustering coefficient with the size of the system N is given by:

$$\langle C \rangle_{ER} = p = \langle k \rangle / N \tag{2.11}$$

and the average path length, on the other hand shows a dependence given by:

$$\langle L \rangle_{ER} \sim \frac{lnN}{ln\langle k \rangle} \tag{2.12}$$

Notice that the value of the clustering coefficient tends to zero in the limit of large N. It is also important to point out that this model produces homogeneous random graphs, which do not share certain topological features with the real networks, for example, they have low values of the clustering coefficient, and do not show any kind of correlations between nodes.

Small-World Networks

A graph in which, although most pairs of nodes are not directly connected to each other, they can nonetheless be in touch by a small number of steps is called Small-world network, since it captures this so-called phenomenon of strangers being linked by a mutual acquaintance (also known as *six degrees of separation* [16–18]). Some properties of real networks can be well modeled using Small-world networks, for example social networks, gene networks or the Internet. Nonetheless, it is important to keep in mind that 'small-world' is a concept that includes several kind of topologies: empirical data [19] suggest the existence of three classes of small-world networks, as far as its degree distribution is concern: scale-free networks, broad-scale or truncated scale-free networks, and single-scale or random networks.

The first Small-world network model was proposed by Watts and Strogatz [15], and it interpolates between a regular graph and a random graph, depending on a single parameter $p \in [0, 1]$, without altering neither the number of nodes nor the number of connections per node of the original graph. This is a random graph generation model that produces networks with Small-world properties, possessing short average path length and high clustering coefficient provided the adequate range of the parameter p (see Fig. 2.8).

Departing from a one-dimensional regular lattice or a ring, where each node has exactly the same number of neighbors, z, we rewire every link with a probability p, avoiding multiple connexions between two nodes and self-connections. In another version of the model, we depart from a ring, where each node has exactly z neighbors, and we add a link between every pair of nodes, with probability p, instead of rewiring the existing links. Regarding the degree distribution, for $p = 0$ we have $P(k) = \delta(k - z)$, where z is the coordination number of the lattice ($z = 4$ in the case shown in Fig. 2.7); whereas for finite values of $p \in (0, 1]$, $P(k)$ still has a peak around z, but it obviously gets broader as p increases. For the cases where $p \in (0, 1]$, the probability of finding a node with a large connectivity decreases exponentially with k, as it happen for ER random networks, so vertexes with large connectivity are practically absent as well. For $p = 0$ we keep the initial ring structure, which has high values both for the clustering coefficient ($C \sim 3/4$), but also for the average path length ($L \sim N/(2k) \gg 1$).

On the other hand, for $p = 1$ we have a random network -though, to be rigorous, in the second version, there are not any nodes with connectivity $k < z/2$, as there would

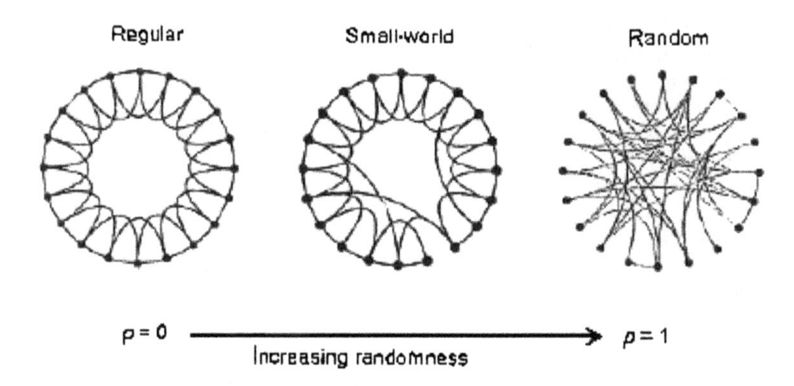

Fig. 2.7 Diagram of the random rewiring procedure for interpolating between a one-dimensional lattice and a random network in the Small-world model. The networks have $N = 20$ nodes and $k = 4$. Original figure from [15]

be in a random network built with a mechanism such as ER. Its average path length is short ($L \approx L_{random} \sim \frac{\ln N}{\ln k}$), but its value for the clustering coefficient is also low ($C \approx C_{random} \sim k/n \ll 1$). Nonetheless, there is an intermediate region of p where we can get a network with both features: a high value for the clustering coefficient and a short average path length. This is due to the presence of long-range connections or shortcuts introduced by the rewiring procedure. Notice that the introduction of these shortcuts makes the average path length drop, not only for the pair of nodes involved, but for all their neighbors too. Moreover, the removal of some links from a neighborhood due to the rewiring process, does not affect the clustering coefficient too drastically, so it remains unaltered for small values of $p \lesssim 0.01$ (see Fig. 2.8). In other words, during the dropping of $L(p)/L(0)$, the clustering $C(p)/C(0)$ remains almost unaltered, which means that this transition to the Small-world is undetectable on a local level.

Regarding the dependence of the small-world behavior with the size of the system, it has been shown [20] that the emergence of this regime occurs for a value of p that approaches zero as N diverges.

The BA Model

Both the Small-world model and the ER model, explained previously, although are most undoubtedly useful and insightful, display two important features that make them very different from the real networks. The first one is the assumption that the whole system is present from the very beginning, it is to say, that the network has a fixed size N and it does not grow, no new nodes are added. In contrast, it has been observed that most real networks are open systems, and they get new vertexes that connect with the ones already present, so the number N keeps increasing throughout the lifetime of the graph. The second one is the assumption that the probability

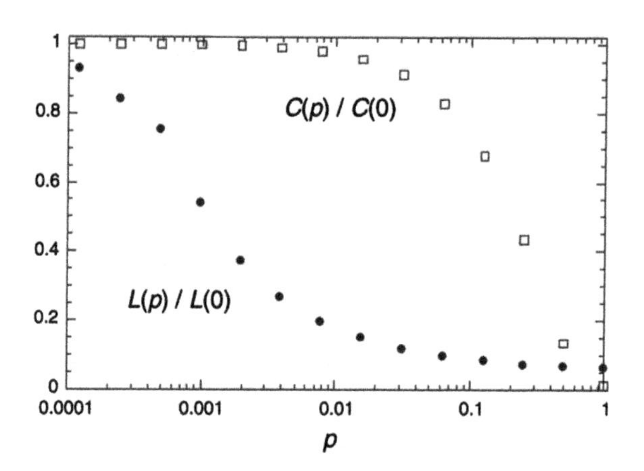

Fig. 2.8 Average path length and Clustering coefficient for the Small-world model, as a function of the probability of rewiring p, normalized by their respective values for the ring, i.e. when $p = 0$. Notice that the x-axis is shown in logarithmic scale. The graphs have $N = 10^3$ nodes and $\langle k \rangle$. The data shown is the average over 20 different rewiring procedures. Original figure from [15]

that two vertexes are connected is uniform. Again, in contrast, most real networks show clearly a preferential attachment: usually, the more connected a node is, the more easily it will get even more neighbors due to connections from new nodes.

The Barabási-Albert (BA) [20] is a model for building scale-free networks that is based on two fundamental ingredients: *preferential attachment*, i.e. the assumption that the likelihood of receiving new edges increases with the node's degree, and *growth*. Actually, variants of the model, with just one of the two ingredients have been tried, but neither of them get networks with power-law distributions. This was a model originally inspired on the growth of the World Wide Web, and as we have already mentioned, the idea behind it is that the highly connected nodes get new links at a higher rate than the lower connected ones or, in other words, the catchphrase '*rich get richer*' [21]. This is a phenomenon easily found on real systems (and it is known in sociology as the Matthew effect [22]).

Thus, we start with a little core of m_0 disconnected nodes, and at each time step $t = 1, 2, 3, \ldots, N - m_0$, a new node i is added to the system with $m \leq m_0$ links to existing nodes. The probability that an existing node j gets one of the links from the newcomer is proportional to its own connectivity, k_j, in a linear way:

$$\Pi_j = \frac{k_j}{\sum_l k_l} \tag{2.13}$$

Since every new node links to m other nodes, at any given moment t, the network has $N(t) = m_0 + t$ nodes and $K(t) = mt$ links. Besides, for long times, the average degree of the network is $\langle k \rangle = 2m$. The degree distribution of these networks is a power law, $P(k) \sim k^{-\gamma}$, with $\gamma = 3$. A scale-free degree distribution implies that there are a lot of nodes with just a few connections, and a small number of nodes with a very high connectivity. These highly connected nodes are called *hubs* and they usually play an important role in most dynamical processes that can take place on the system, as we will see with some detail during this thesis. Besides, the degree distribution $P(k)$ of the BA networks is independent of time, and thus independent of the size of the system, indicating that despite its continuous growth, the system organizes itself into a scale-free stationary state.

The dependence of the clustering coefficient with the size of the system N is approximately a power law, given by:

$$\langle C \rangle_{BA} \sim N^{-0.75} \tag{2.14}$$

The average path length, on the other hand shows a dependence given by:

$$\langle L \rangle_{BA} \sim \frac{lnN}{ln(lnN)}. \tag{2.15}$$

The value of the average path length in BA networks is smaller than in ER networks for any value of N, so obviously, the heterogeneous topologies help bringing the nodes together more than the homogeneous ones. On the other, hand, comparing the values for the clustering coefficient, the corresponding values for the BA networks are about five times higher than for ER networks, and this factor even increases slightly with the size of the system. Moreover, it is worth pointing out the existence of the so-called *age correlations* [23–25] among nodes for the scale-free topologies, which means that the older nodes, i.e. the ones that appear first on the system, are more likely to end up being hubs, just by construction, while the later a node appears, the lower connectivity it will get.

We consider that it is important to stress again that SF networks built via this BA procedure have very low values for the clustering coefficient, when comparing with real networks, so we must admit that this kind of topologies might reproduce the degree distribution of those systems, but can not do the same for the clustering coefficient. Along these lines, there have been some other models that, based on BA, tried to put a remedy to this fact. For example, the work by Holme and Kim [26], presents a model for constructing SF networks with tunable clustering coefficient. In few words, this model starts with a set of m_0 unconnected nodes and adds a new one to it every time step, up to N. Each one of the new nodes launches $m \leq m_0$ links. The probability of an existing node i to receive the first link of a newcomer j is proportional to its connectivity k_i, but for the remaining $m - 1$ links that the new node j has to establish, there is a probability p to launch them to a (randomly selected) neighbor of i, and a probability $(1-p)$ to launch them following the original preferential attachment rule. In this way, the family of networks we obtain have all exactly the same power-law degree distribution $P(k) \sim k^{-3}$, but the higher the value of the probability p, the higher the value of the clustering coefficient (it can easily achieve values of 0.5, when we recall that for BA, it tends to zero as N increases, so the order of magnitude of a typical value can be around 10^{-2} for $N = 10^3$). For the particular case $p = 0$, we recover the original BA model, obviously. Moreover, with this Holme-Kim model, the clustering coefficient is independent of the size of the system, as opposed to what happens with BA, where it decreases with N, as we have seen. On the other hand, it is also worth mentioning that, one may think that by increasing p the average path length of the final structure will decrease, since some links that would help shortening it by linking to nodes far apart, are now linking nodes in the same neighborhood. As it turns out, the value of the average path increases slightly with the probability p, but the dependence with the size of the system remains logarithmic, so we do not lose the 'small-world' property with this model.

Finally, we also want to remark two points regarding preferential attachment. First, other mechanisms for building SF networks have been proposed [27], that are not based on growth and preferential attachment like the BA model is. Instead, an intrinsic fitness (from a given probability distribution) is assigned to each node in the system, and then pairs of them are linked together, according to a function of their fitness. And second, if one combines growth, preferential attachment and some aging mechanism or introduces a cost per link, then one will obtain SF topologies with a cutoff on the degree distribution, or even make the scale-free regime disappears altogether [19].

The GM Model

The Gardeñes-Moreno is a model [28] that interpolates between Erdös-Rényi random networks and Barabási-Albert scale-free networks as far as the degree distribution is concerned, through a tunable parameter α, so it generates a one-parameter family of networks. This parameter $\alpha \in [0, 1]$ determines the degree of heterogeneity of the network, whose final size will be Ω. Thus, $\alpha = 0$ gives rise to scale-free networks and $\alpha = 1$ to random graphs, and for in-between values, the topology will have an intermediate degree of heterogeneity.

The procedure to generate these networks is as follows: we start with a small fully connected core of m_0 nodes, and a set $\mathcal{U}(0)$ of $(\Omega - m_0)$ disconnected nodes. At each time step, a new node j from the set $\mathcal{U}(0)$ is chosen, and it makes a link in two possible ways: with a probability α, it attaches to any other node i from the whole set of $\Omega - 1$ nodes with uniform probability:

$$\Pi_i^{uniform} = \frac{1}{\Omega - 1} \tag{2.16}$$

and with probability $1 - \alpha$, it establishes a link following a preferential attachment (PA) strategy. This means that the probability for any other node i to get attached to node j is a function of its connectivity, in a way given by:

$$\Pi_i^{PA} = \frac{\hat{k}_i^{pa} + A_i}{\sum_{l \in \Omega} (\hat{k}_l^{pa} + A_l)} \tag{2.17}$$

where $\hat{k}_i^{pa}$ is the incoming PA degree of the node i, that is, those links received by i when other node launches (in average) $(1 - \alpha)m$ links following the PA rule. On

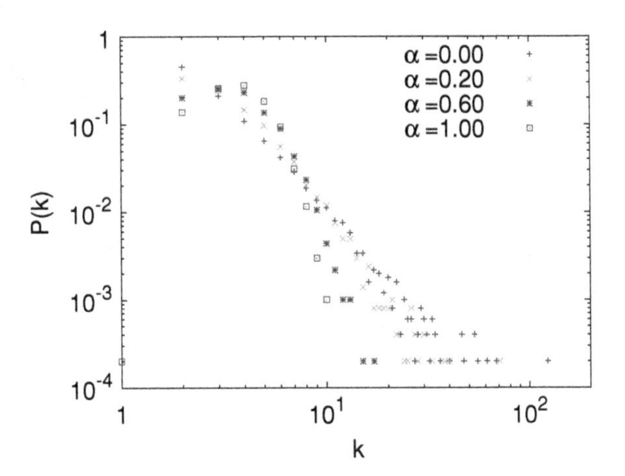

Fig. 2.9 Degree distributions for several networks obtained for the shown values of the parameter α with the GM model, interpolating between random ($\alpha = 1.0$) and scale-free ($\alpha = 0.0$) graphs. The size of the system is $\Omega = 5 \cdot 10^3$ and $\langle k \rangle = 2m = 4$. Every point is the average over 10^3 different realizations

the other hand, A_i is an initial attractiveness (or fitness) the new node has when it is introduced in the connected component (either because it is chosen at random by any node or because it is launching its m outgoing links over the rest of nodes). This associated parameter is zero if the node i is not in the connected set and is $A_i = A$ if it is linked to other nodes, i.e., if it belongs to $N(t)$. Thus, the preferential attachment is strongly correlated with the simultaneous uniform random linking, and, on the other hand, it is linear with the incoming PA degree of the node $\hat{k}_i^{pa}$. Next, we repeat the linking procedure for another $m - 1$ times for the same node j, and then we repeat the whole process altogether for the rest of the nodes, i.e., for another $\mathcal{U} = \Omega - m_0$ more time steps.

On Fig. 2.9 we show the degree distribution for some networks obtained with the GM model, for several values of the parameter α but the same size Ω and average connectivity k. Notice that the transition between heterogeneous and homogeneous topologies is smooth, as α increases.

2.1.4 Processes on Networks

So far in this chapter, we have studied some general topological properties of networks, as well as some well known and widely used models to generate them, and some real examples too. Nonetheless, we have to keep in mind that the ultimate goal of studying these structures, is to finally be able to model, describe and predict the different dynamics that can take place on top of them. Those include a wide and varied collection, such as disease [1, 2, 29–35] or rumor spreading, synchronization [1, 36–40], diffusion, traffic information and congestion, network search and navigation, percolation, robustness against random failures or targeted attacks [41, 42], cultural dissemination, opinion formation or language dynamics [43], and games [44]. In this section, it is not our intention to go exhaustively though all of them at

Fig. 2.10 Schematic representation of the SIR model

all (for some very nice reviews on the subject, see [1, 2, 40, 44]), but just to briefly examine a few of them, as an example, describing some the most popular models or approaches that have been proposed, and also pointing out the differences introduced by the underlying topology on the outcome of the dynamics, in comparison to well-mixed situations or lattices.

Disease Spreading

Epidemic spreading is a very interesting and obviously very important object of study [1, 2, 29–35]. The aim in this field is not only to understand the mechanisms through which diseases spread on a population, but also to design strategies to control them, and to be able to protect the population from pandemics.

Specifically, *Compartmental Models* in epidemiology stand for some models that, in order to describe the progress of an epidemic in a large population comprising many different individuals, reduce such population diversity to a few key characteristics which are relevant to the infection under consideration. For example, for most common childhood diseases that confer long-lasting immunity, such as the chickenpox, it makes sense to divide the population into those who are susceptible to the disease, those who are infected and those who have recovered and are therefore immune. Thus, one can ignore the rest of the information about the population, such as age distribution or race, because it is irrelevant for the model. These subdivisions of the population are called *compartments*.

In particular, one of the more used (and at the same time simple) models to study disease spreading is the SIR model. It considers that the population is compartmentalized into three possible states: Susceptible, Infected (and infectious), and Recovered (or removed). Thus, a susceptible individual can get infected with a certain probability if it is in direct contact with an infected one, and in turn, an infected individual recovers (or dies) with a different probability, not being able to get infected again in any case. This simple model describes many infectious diseases, such as measles, mumps and rubella. On Fig. 2.10 we show a simple scheme for the dynamics of this model. Of course, there are other models much more sophisticated, that take into account other intermediate states in the infectious process, such as latency, infected asymptomatic individuals or vaccination (see for example [44, 45]).

As a first approximation, one can consider the homogeneous mixing hypothesis, which assumes that people with whom a susceptible individual has contact are chosen at random from the whole population. This is a strong and somehow questionable assumption, since it does not take into account local details, such as individual diversity on the number of acquaintances, community structure or geographic constrictions. And, on the other hand, one should take into account that some illness like the common cold, can be modeled accurately enough as a random-contact process, ignoring the social structure underneath, while it has been proved than for some others, such as the venereal diseases, one can not even describe them using a random degree distribution for the population, but a scale-free, so in these cases, the structure is essential.

Nonetheless, this approximation made by the SIR model allows us to describe analytically the behavior of the models simply by using ordinary differential equations for the densities of individuals in each compartment:

$$\begin{aligned}\frac{ds(t)}{dt} &= -\lambda\bar{k}\rho(t)s(t)\\ \frac{d\rho(t)}{dt} &= -\mu\rho(t) + \lambda\bar{k}\rho(t)s(t)\\ \frac{dr(t)}{dt} &= \mu\rho(t),\end{aligned} \tag{2.18}$$

where $s(t)$, $\rho(t)$ and $r(t)$ are respectively, the fraction of susceptible, infected and recovered individuals on the population at time t, so $s(t) + \rho(t) + r(t) = 1$. On the other hand, one susceptible individual becomes infected (if in contact with another infected one) with a probability λ, an infected individual recovers (or dies) with a probability μ, and $\bar{k}$ stands for the connectivity of the population, assumed exactly the same for everyone.

The most relevant prediction of this model is the existence of a non-zero *epidemic threshold*,

$$\lambda_c = 1/\bar{k} \tag{2.19}$$

so if $\lambda > \lambda_c$, the disease spreads and infects a finite fraction of the population, and if $\lambda < \lambda_c$, the total number of infected individuals (the so-called *epidemic incidence*, defined as $r_\infty = lim_{t\to\infty}r(t)$) is infinitesimally small in the limit of a large population.

On the left panel of Fig. 2.11 we show an example of time evolution of the dynamics for a meaningful set of the parameters, namely, for $\lambda = 0.94$, $\mu = 1.0$, $\bar{k} = 6$ and using as inital conditions: $s(0) \simeq 1$, $\rho(0) \simeq 0$ and $r(0) \simeq 0$. On the right panel, it is shown the dependence of the epidemic incidence with the infection probability λ.

To deal with situations where the population is not well-mixed, or as we have mentioned before, the nature of the disease itself does not allow us to treat the pattern of interactions as homogeneous, we will need to represent the system as a graph, where nodes are the individuals (belonging to one of the three possible states: Susceptible, Infected or Recovered), and links are the interactions through which a susceptible node can become infected, if it has another infected node as a neighbor. So now, we want study the SIR process on an uncorrelated heterogeneous network (with generic degree distribution $P(k)$ and a finite average connectivity $\langle k\rangle$). We will study $s_k(t)$, $\rho_k(t)$ and $r_k(t)$, meaning the time evolution of the fractions of susceptible, infected and recovered individuals, respectively, within a connectivity class k, and with the normalization condition: $s_k(t)+\rho_k(t)+r_k(t) = 1$ for any given connectivity

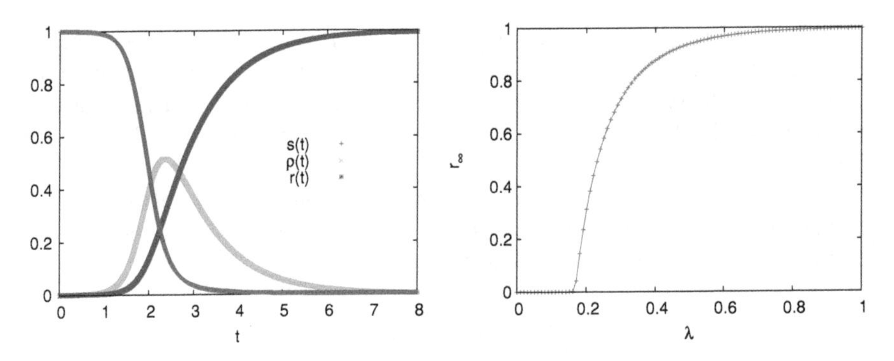

Fig. 2.11 Time evolution of the SIR dynamics (*left*) for $\lambda = 0.94$, $\mu = 1.0$, $\bar{k} = 6$ and taking $s(0) \simeq 1, \rho(0) \simeq 0$ and $r(0) \simeq 0$ as initial conditions, and the dependence of the epidemic incidence (*right*) with the probability of infection, λ for $\mu = 1.0$ and $\bar{k} = 6$

class and time instant. The global magnitudes are now given by the average over all the classes of connectivity present on the graph, so for example, the total fraction of infected individuals on the population at a given time t is: $\rho_k(t) = \sum_k P(k)\rho_k(t)$. Here it is important to notice that the network is considered static, so $P(k)$ does not change over time.

The equations for the evolution of the three compartments are similar to Eq. 2.18, but now we differentiate among connectivity classes:

$$\begin{aligned}\frac{ds_k(t)}{dt} &= -\lambda k s_k(t)\Theta(t)\\ \frac{d\rho_k(t)}{dt} &= -\mu\rho_k(t) + \lambda k s_k(t)\Theta(t)\\ \frac{dr_k(t)}{dt} &= \mu\rho_k(t),\end{aligned} \tag{2.20}$$

where $\Theta(t)$ is the probability of a given link to point towards an infected node, and is given by:

$$\Theta(t) = \frac{\sum_k k P(k)\rho_k(t)}{\langle k \rangle}. \tag{2.21}$$

Notice that this probability is the same for any node we consider, so it does not take into account any possible correlations between the connectivity of the nodes.

Again, one can get that there is an epidemic threshold, given by:

$$\lambda_c = \frac{\langle k \rangle}{\langle k^2 \rangle} \tag{2.22}$$

below which the epidemic incidence is zero, and above which it has a finite value. As we can see, this threshold depends inversely on the connectivity fluctuations of the network the disease is spreading on, so for a system whose topology has a finite value, $\langle k^2 \rangle$, such as a random graph, then we get a threshold with a finite value as well (and therefore, a standard phase transition scenario). However, for scale-free networks we know that their connectivity fluctuations $\langle k^2 \rangle$ diverge when $N \to \infty$, which implies a vanishing epidemic threshold for increasingly larger systems.

The absence of a threshold in scale-free topologies is an important result that differs drastically from the one obtained for random networks or well-mixed scenarios, and it should be taken into account, for instance for prevention or vaccination strategies to be used by the health authorities, in order to efficiently fight off epidemics.

On the other hand, it is also worth noticing that real networks, even when they present some degree of heterogeneity on the connections, do have a finite size, and thus an effective threshold, depending on its $\langle k \rangle$ and $\langle k^2 \rangle$. Nonetheless, this value is usually very small for a large enough population, and is considerably smaller than the one for a random graph of the same size.

With regard to immunization strategies on scale-free topologies, we can point out that random vaccination is not effective, since there is always a non-zero epidemic incidence, even for very high vaccination ratios among the population. Nonetheless, targeted immunization, i.e., vaccinating the most connected individuals in a population, can give better results. On the other hand, is not always realistic to assume that the number of connections of a node on a real network can be known. A possible solution to this problem is the vaccination of random acquaintances of random chosen individuals, since the probability of reaching a particular node by following a randomly chosen edge is proportional to its degree.

Finally, we will say that for correlated networks it has been found that the qualitative behavior is the same as for uncorrelated networks, although there are some quantitative differences: on the one hand, while the likelihood of an epidemic outbreak is not modified when taking into account positive correlations, the epidemic incidence is smaller than in networks without correlations, and on the other hand, the diseases can live longer in assortative topologies.

Synchronization

Synchronization [1, 36–40] is a self-organized phenomenon where a set of individuals, initially acting on their own, gradually become more similar in their deeds, without any appointed leader or environmental external signal to guide them. In this way, after some time, they start behaving under the same pattern, showing, if not total, at least some identifiable level of clocking: they became somewhat 'in sync'. There are many examples of synchronization in natural and human systems: crickets chirping in a summer night, neurons firing at the same pace, kids playing or singing along on spur of the moment, or groups of women living together, whose periods synchronize,...

A simple model has been frequently used in order to address synchronization: the Kuramoto model. It approaches the problem considering a mean field approximation, where every individual is an oscillator, and they are all supposed to interact to everyone else through a purely sinusoidal coupling, so the governing equations for each one of them is given by:

$$\dot{\theta}_i = \omega_i + \frac{K}{N}\sum_{j=1}^{N}\sin(\theta_j - \theta_i) \tag{2.23}$$

where K is the coupling constant, ω_i is the natural frequency of the oscillator i, and the factor $1/N$ is incorporated to make sure that the system behaves correctly in the thermodynamic limit. The natural frequencies are assumed to be distributed according to some unimodal and symmetric function, whose mean frequency is Ω.

The collective behavior of the whole system is described by the macroscopic complex order parameter:

$$r(t)e^{i\phi(t)} = \frac{1}{N}\sum_{j=1}^{N} e^{i\theta_j(t)} \tag{2.24}$$

so the modulus $0 \leq r \leq 1$ measures the *phase coherence* of the population, whereas $\phi(t)$ is the average phase. The value $r \simeq 0$ corresponds to the lack of synchronization (the oscillators move incoherently) and $r \simeq 1$ to the case where almost the whole system is in sync (their phases are locked). The existence of a critical value, K_c, can be derived for the coupling, which separates a 'disordered' from an 'ordered' regime. In this second regime (when $K \geq K_c$), there are two types of long term behavior: a group of oscillators for which $|\omega_i| \leq Kr$, that are phase-locked at frequency Ω, and the rest of them, with $|\omega_i| > Kr$, that are drifting around the circle, sometimes accelerating and sometimes rotating at lower frequencies.

If one should include some kind of structure in the population in order to give an account of the complex interaction patterns among individuals, then, instead of Eq. 2.23, one needs to consider an extension of it:

$$\dot{\theta}_i = \omega_i + \sum_{j=1}^{N}\sigma_{ij}a_{ij}\sin(\theta_j - \theta_i) \tag{2.25}$$

where σ_{ij} accounts for the specific coupling strength between individuals i and j, and a_{ij} is the adjacency matrix of the network.

The mean field approach for complex networks considers that every oscillator is influenced by the local field created in its neighborhood, so the local order parameter is proportional to the connectivity of the node, k_i. It can be obtained the critical coupling for this situation:

$$\sigma_c = K_c\frac{\langle k\rangle}{\langle k^2\rangle}. \tag{2.26}$$

It is to say, we get a rescaled critical value for the all-to-all topology, K_c, by the ratio between the mean connectivity of the particular network and its fluctuations.

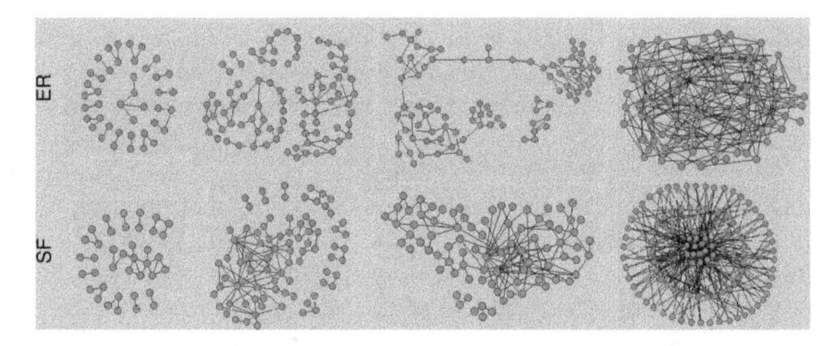

Fig. 2.12 Squematic representation of the different paths to synchronization displayed for SF (*bottom*) and ER (*top*) networks (higher values of the coupling strength are shown from *left* to *right*. Original figure from [36]

So once again, it is clear that for random networks there will be a threshold, but for (infinite) SF networks, this critical value will tend to zero.

Besides, it is important to point out that no exact analytical results for the Kuramoto model on general complex networks are available up to date, but one can always numerically simulate its dynamics. These simulations [36, 39] confirm the theoretical predictions, since they have shown that the onset of synchronization first occurs for SF, and as the topology becomes more homogeneous, the critical point moves to larger values, and the system seems to be less synchronizable. On the other hand, the particular paths to synchronization [36, 40] are also very different depending on the underlying structure (see Fig. 2.12): in SF networks, links and nodes are incorporated together to the largest of the synchronized clusters, while for homogeneous topologies, what are added are links between nodes already belonging to such cluster, making the route to complete synchronization a 'sharper' process, somehow. In other words, in the presence of hubs, a giant component of synchronized pair of oscillators forms and grows by recruiting nodes linked to them, while on the contrary, in homogeneous structures, many small clusters first appear and then group together.

Cultural Dissemination

A very interesting aspect of human interactions is how people from different cultures can relate to each other, changing some of their own cultural traits when they meet. Nonetheless, if two individuals do not share any cultural features to begin with, it will be probably very hard for them to communicate and interact, but if they do have initially something in common (like some interests, hobbies, goals or even an aversion against something), they may start some kind of relationship. Moreover, it makes sense to assume that the more similar they are before meeting each other, the more likely it is for them to interact and become even more similar after that (this phenomenon is known as *homophilia*). As a result, not only individuals, but also societies change over time due to this mechanism of cultural influence. Thus, one would expect these societies to became homogeneous (global) over time, as far as culture is concern, but as it turns out, sometimes they do not. Instead, such interactions can give rise to different groups with practically nothing in common, surprisingly enough.

Since Axelrod proposed his agent-based model [46] to address the issue of cultural dissemination in 1997, much effort has been put on studying these kind of processes [43, 47–52]. Under this paradigm, we generally consider that an individual's culture can be represented in terms of a set of attributes, such as language, religion, technology, dressing style, literary and musical preferences, sport preferences, and so on. Thus, an individual can be represented using a vector $\vec{V_i} = (v_i^1, v_i^2, \ldots, v_i^F)$, with $i = 1, 2, \ldots, N$, and where F is the total number of features that define a culture. Each one of these components can take only Q integer values, or cultural traits, and we assume that Q is the same for the F features. It is worth noticing that within this model, we do not consider as 'cultural' those features an individual can not change, for example skin color or physical constitution. Besides, we will consider our society to be placed in a lattice of size $L \times L = N$, where individuals will interact only with their neighbors.

Once we have randomly distributed the initial values for all the features of every individual in the system, the cultural interaction dynamics is defined as follows: every time step, an individual i is randomly chosen and one of its neighbors j, is also randomly selected. One measures the overlap between their cultural vectors, given by:

$$S_{ij} = \frac{1}{F} \sum_{l=1}^{F} \delta(v_i^l - v_j^l) \tag{2.27}$$

where $\delta(x) = 1$ if $x = 0$ and $\delta(x) = 1$ otherwise. If these two individuals are totally different ($S_{ij} = 0$) or exactly the same ($S_{ij} = 1$), then nothing happens, since the link between them is *blocked*. But if that is not the case, and $S_{ij} \in (0, 1)$, then the link is '*active*', and we then consider the value of the overlap S_{ij} as the probability that one of them imitates the other in one of the other features they have different. Obviously, the more similar they are, the higher the probability of becoming even closer through social interaction.

Letting the system evolve, it will eventually reach a *frozen state*, meaning that all the links between individuals are blocked. A useful order parameter is the relative size of the largest cultural cluster, $S_{\max}$, it is to say, the largest group of individuals that share the values for all their cultural features. According to some studies on lattices [47, 49, 51, 53], when $F > 2$, a non equilibrium first-order phase transition from order to disorder is observed as a function of the number of traits Q (the control parameter). There is a critical value, so if $Q < Q_c$, the final state of the system corresponds to $S_{\max} \sim 1$, a *global*, homogeneous state, while if $Q > Q_c$, then $S_{\max} \ll 1$, a *polarized* state with different *cultural domains* arises (see Fig. 2.13 (**left**)). This transition gets sharper as the size of the system increases.

If we analyze the time evolution of the relative number of blocked links (see Fig. 2.13 (**right**)), it can be seen that there is a non-zero initial value, due to just

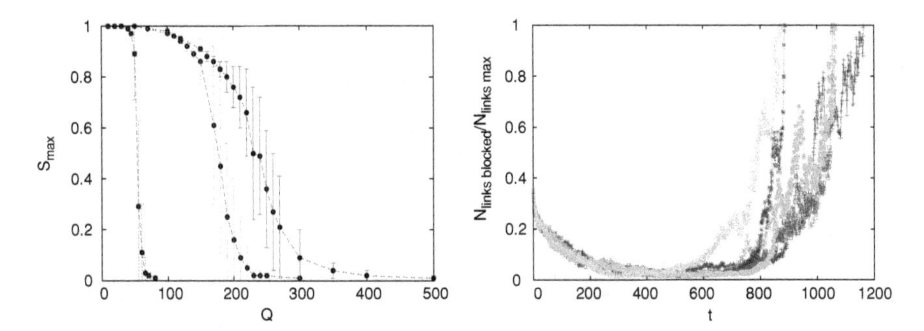

Fig. 2.13 *Left* Dependence of the largest cluster of global cultural consensus with the number of traits per feature for a 50×50 node square lattice with a 4-node neighborhood (*left*), ER random network (*center*) and BA scale-free network (*right*), and always for $F = 10$. The last two topologies have $\langle k \rangle = 6$ and $N = 10^3$ nodes. Every point is the average of 100 independent realizations. *Right* Several examples of time evolution of the relative number of blocked links. The underlying topology is a SF network made up of $N = 10^3$ nodes and $\langle k \rangle = 6$ and for a fixed value of $F = 10$

random assignment of the traits, that drops quickly as the dynamics starts, and individuals begin to interact. Then, this magnitude remains very low for a considerable amount of time, to finally rise up to the final value, corresponding with the rapid rise of $S_{\max}(t)$. This reflects the fact that, while the individuals have almost nothing in common, the system seems to spend a lot of time in that state, unable to get to an agreement, but once the individuals share some values for the features, then the final state is rapidly achieved. Notice that every realization shown in Fig. 2.13 (**right**) reaches its final state at its particular 'consensus time', since it is an stochastic process.

If we consider now that the pattern of interactions is given by a finite complex network [47], instead of by a lattice, the general picture of the phase transition remains unaltered (see Fig. 2.13 (**left**)), but with a higher value for Q_c (even higher for SF than for random networks, but qualitatively similar).

On the other hand, recent studies [52] have shown that, one can analyze the cultural evolution process towards the final state, from a global point of view (it is to say, considering the macroscopic level of consensus in the system though $S_{\max}$), but also from a feature level. It means that at any given time, we consider F layers or subgraphs of the original graph G. In the subgraph $G_f(t)$, two individuals are connected if they are physically connected in G, and if they share the value of the feature f at that precise instant of time. In this way, we can observe how cultural consensus evolve in every layer, $S^f_{\max}$, and we get to discover that there are some relevant differences between the two approaches: while for the global consensus point of view, the system remains apparently unordered for a large fraction of the simulation time, to finally get organized very quickly (Fig. 2.14 (**left**)), the organization at a feature level starts much earlier. Actually, $S^f_{\max}$ increases monotonously over time from the very beginning (Fig. 2.14 (**right**)).

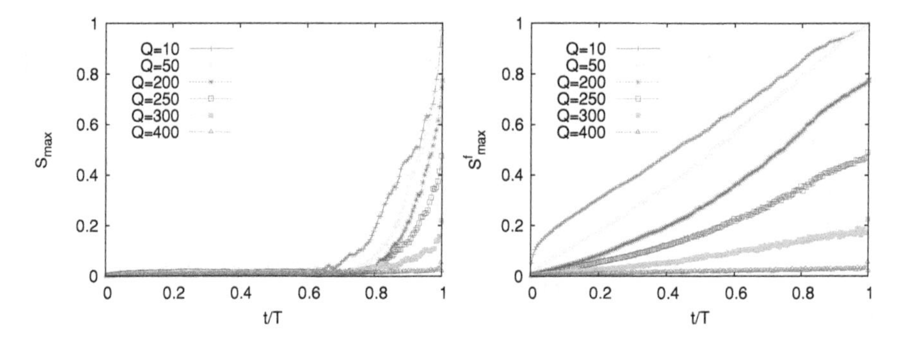

Fig. 2.14 Time evolution (relative to the final consensus time T) of the largest cluster of cultural consensus at global (*left*) and at feature (*right*) level for a value of $F = 10$ in SF networks made up of $N = 4 \cdot 10^3$, with average connectivity $\langle k \rangle = 6$

Finally, it is also worth mentioning that there are many other works with different variations of the Axelrod model [43], including for example noise [54], an external field [55], rewiring of the connections between nodes [56], even movility of the individuals [57], or even a combination of the original Axelrod model for cultural dissemination with the original Schelling model of social segregation [58].

2.2 Games

A *game* can be considered as a formal abstraction of social interactions between individuals. There must be at least two decision makers (or *players*), who can choose between at least two different actions (also called *strategies*). It is worth stressing that a player does not need a brain in order to adopt a strategy, on the contrary, they can be very simple agents: bacteria, for example, have the basic capacities to play games, since they are highly responsive to certain aspects of their chemical environment, and they can respond differently depending on the actions of their neighbors, the behavior can affect the fitness of others and vice versa, and finally, the conditional strategies can be inherited by the offspring [59]. The outcome of the interaction depends on the strategy every player adopts. Thus, *Game Theory* is a branch of applied Mathematics that tries to capture these situations and it is usually considered to have its origin with the work of von Neumann and Morgenstern [60] in 1944. Historically, Game Theory has been used in very different fields, such as economics, biology, political science or sociology, and there are two main different approaches: Classic Game Theory and Evolutionary Game Theory, which made different assumptions about the systems they model.

Classic Game Theory formally studies how rational players should behave in order to obtain the maximum possible benefit or payoff. Nonetheless, one could easily object to the concept of 'rational player' as an accurate representation of real individuals in a social or biological context. 'Rational player' means that its only goal and motivation is to maximize its benefits, given its belief about its opponent's strategy, but there are plenty of real situations where the actions of the players do not seem to aim a maximum payoff.

Evolutionary Game Theory [61–63] was originated in 1973 by the work of Maynard Smith and Price works. It studies the time evolution of large populations of individuals who repeatedly play a game and are exposed to selection and replication (with or without mutation). Their strategies are fixed, and usually, the encounters between the individuals are supposed to happen at random, in a 'well-mixed' situation, so there is no social structure behind it (everyone interacts with everyone else), and it allows for the analytical treatment of the problem. Thus, the probability of interacting with an individual that uses strategy i is proportional to the fraction of individuals that are using that particular strategy in the system at the moment, x_i. The payoffs from all these interactions are added up, and success in the game is interpreted as reproductive success. Thus, payoff means fitness in the Darwinian way: the strategies that perform better, reproduce faster, which can be straightforwardly interpreted as natural selection.

In this section we intend to establish just a few useful concepts and results in Classical Game Theory, always keeping in mind that our goal is to understand the problem of cooperation. Then we will move on to the approach given by Evolutionary Game Theory, and finally, we will point out some mechanisms that have been introduced to explain the survival of cooperation observed in several natural and social systems, specially, the differences in the outcome of a game when dealing with a structured population, it is to say, when we have an underlying topology.

2.2.1 Classical Game Theory

In Classical Game Theory (CGT), we consider that interacting individuals can choose a strategy -or a way to act- among a well-defined set of them. A game is called *normal-form* if it is determined by a payoff matrix. Thus, for instance in a 2×2 game, we have two players and two different strategies A and B, and then depending on their particular choices, the benefits the players will obtain are given by the payoff matrix:

$$\begin{array}{c c} & \begin{array}{cc} A & B \end{array} \\ \begin{array}{c} A \\ B \end{array} & \begin{pmatrix} a & b \\ c & d \end{pmatrix} \end{array} \tag{2.28}$$

This means that, for instance, when a player uses strategy A against a player using also A, it get a payoff equal to a, when a player uses strategy A against a player using a strategy B, it get a payoff equal to b, and so on. We say that strategy A *dominates* strategy B, if $a > c$ and $b > d$. In that situation, no matter what strategy your opponent uses, it is better always to use A. Conversely, B *dominates* A, if $a < c$ and $b < d$.

Now, in a general case of a $N \times N$ payoff matrix U, if we denote the N *pure strategies* by $R_1, R_2, \ldots R_N$, then the simplex S_N of the linear combinations of pure strategies:

$$S_N = \left\{ p = (p_1, p_2, \ldots, p_N) : p_i \geq 0 \;\; and \;\; \sum_i p_i = 1 \right\} \tag{2.29}$$

is the set of *mixed strategies*. A mixed strategy can be seen as the one used by a player that chooses strategy R_i with a probability p_i, where $i = 1, 2, \ldots, N$. The N vertexes of the simplex S_N are the N pure strategies, while the interior of the simplex is the set of completely mixed strategies, it is to say, those for which $p_i > 0 \;\; \forall i$. The boundaries of the simplex, on the other hand, correspond to mixed strategies that must have necessarily one of the probabilities set to zero. We can calculate the benefit of a p-strategist against a q-strategist as:

$$pUq = \sum_{i,j} p_i u_{ij} q_j \tag{2.30}$$

and the set of strategies for which the application $p \rightarrow pUq$ achieves its maximum value is called *best responses* to q.

A strategy q is called a *Nash Equilibrium* (originally called 'equilibrium for n-person games' by Nash in 1950 in [64]) if it is the best response to itself. This means that if two individuals are both using a strategy that is a Nash Equilibrium, then neither of them can unilaterally deviate form that strategy and increase its payoff. Moreover, a Nash Equilibrium is called *Strict* if it is the only best response to itself, therefore $\forall p \neq q$ it is fulfilled that $pUq < qUq$. If q is a Nash Equilibrium, then there is a constant c that satisfies that $(Uq)_i \leq c$, and from this result can be derived that a Nash Equilibrium is always a pure strategy.

A strategy $\hat{p}$ is *Evolutionary Stable* if $\forall p \in S_N$ with $p \neq \hat{p}$ the inequity:

$$pU(\epsilon p + (1 - \epsilon)\hat{p}) < \hat{p}U(\epsilon p + (1 - \epsilon)\hat{p}) \tag{2.31}$$

is fulfilled $\forall \epsilon > 0$, as long as it is smaller than a certain appropriate invasion threshold $\bar{\epsilon}(p)$. It can be proven the following logic chain:

Strict Nash Equilibrium $\rightarrow$ Evolutionary Stable Strategy $\rightarrow$ Nash Equilibrium.

Let's now consider again a particular set of 2×2 games. We can analyze the possible outcomes within the CGT framework. We consider two different strategies: cooperate (C) and defect (D), and the correspondent payoff matrix:

$$\begin{array}{cc} & \begin{array}{cc} C & D \end{array} \\ \begin{array}{c} C \\ D \end{array} & \begin{pmatrix} R & S \\ T & P \end{pmatrix} \end{array} \tag{2.32}$$

Depending on the relative ordering of the parameters, we can define three games:

- The Hawks and Doves (or Snow Drift or Chicken) game [65–68] fulfills: $T > R > S > P$. Players are referred to as greedy, since they prefer unilateral defection to mutual cooperation ($T > R$). In this situation, C is the best response for D, and vice versa, so one should always try to choose the opposite of what the opponent does, in order to maximize the benefits.
- The Stag Hunt game [69, 70] satisfies $R > T > P > S$. Players prefer mutual defection to unilateral cooperation ($S < P$), resulting in an intrinsic fear of individuals to cooperate. In this situation, C is the best response for C, and D is the best response for D, or in other words, both are Nash equilibria, so it is better always to try to play the same strategy as your opponent.
- The Prisoner's Dilemma game [59, 63, 71–73], for which $T > R > P > S$, both tensions described above are incorporated at once, so is the most difficult situation for cooperation to arise. In this scenario, D dominates C. No matter what strategy your opponent uses, it is better always to defect.

2.2.2 Evolutionary Game Theory

Within the Theory of Evolution, the central actor of an evolutionary system is the *replicator*. A replicator is an entity that possesses the ability of making copies of itself. It can be a gene, an organism, a strategy in a game, a particular belief or opinion, a technique or any other cultural trait in general. A *replicator system* is a set of replicators in a particular environment, with some kind of interaction among the individuals. An evolutionary dynamics of a replicator system is a process of change over time on the replicator frequency distribution, in such a way that the strategies with higher benefits reproduce at a faster pace.

Let us consider that the population is divided into n types of individuals E_1, E_2, ..., E_n with frequencies (or relative abundances) $x_1, x_2, \ldots, x_n$ respectively. The fitness (or expected number of descendants) f_i of the type E_i will be assumed to be a function ot the composition of the whole population. If the population is big enough, and the individuals of a generation are supposed to meet and interact continuously and at random (*well-mixed* scenario), then we can consider that the state of the system $x(t)$ evolves in the simplex S_n as a derivable function of time. The increase of the rate $\dot{x_i}/x_i$ of the type E_n is a measure of its success, in the Darwinian evolutionary sense of the term. Then, we can express this success as the difference between the fitness f_i of this type and the average fitness of the population, $\bar{f}(x) = \sum_i x_i f_i(x)$, and thus describe the evolution of every type in the population using the *Replicator Equation* [62, 74–76]:

$$\dot{x_i} = x_i[f_i(x) - \bar{f}(x)] \tag{2.33}$$

with $i = 1, 2, \ldots, n$. It is easy to see that the simplex S_n is invariant under these equations, so if $x(0) \in S_n$, then $x(t) \in S_n \forall t > 0$. Moreover, the faces of the simplex are also invariant: if one or several strategies are not present at a given moment t_0 of the evolution of the system, then they will never be for any t_1, such as

$t_1 > t_0$. In the case of having mixed strategies, we can also obtain the correspondent Replicator Equation. If there is a game with N pure strategies $R_1, R_2, \ldots, R_N$ and a $N \times N$ payoff matrix U, then a strategy is a point in the simplex S_N, and the $E_1, E_2, \ldots, E_n$ types of individuals present in the system correspond to n points $p^1, p^2, \ldots, p^n \in S_N$.

The state of the whole population is given by the frequencies x_i of the types E_i. The benefits of a p^i-strategist playing against a q^i-strategist is given by $a_{ij} = p^i U p^j$, and thus, the fitness f_i of the type E_i is $f_i(x) = \sum_j a_{ij} x_j = (Ax)_i$. A state $\hat{x} \in S_n$ is a Nash Equilibrium if $xA\hat{x} \leq \hat{x}A\hat{x}$, $\forall x \in S_n$, and it can be proven that if $\hat{x}$ is a Nash Equilibrium, then it is an equilibrium point of the Replicator Equation. A state $\hat{x} \in S_n$ is said *evolutionary stable* if $\forall x \neq \hat{x}$ in an environment of $\hat{x}$ it is fulfilled that $\hat{x}Ax > xAx$. The same way, it can be proven that if $\hat{s}$ is an evolutionary stable state, then it is a point of asymptotically stable equilibrium of the Replicator Equation (but the reciprocal result is not necessarily true).

Replicator Equation for 2 x 2 Games

For the particular case of a 2×2 symmetric game, we will have again that the generic payoff matrix is given by:

$$\begin{array}{cc} & \begin{array}{cc} A & B \end{array} \\ \begin{array}{c} A \\ B \end{array} & \begin{pmatrix} a & b \\ c & d \end{pmatrix} \end{array} \tag{2.34}$$

And according to the Evolutionary Game Theory, we should consider that the fitness of an individual playing a certain strategy depends on the fraction of individuals that play every strategy (it is to say, the so-called *frequency-dependent selection*), so if the vector $\vec{x} = (x_A, x_B)$ represents the composition of the population, in terms of the two possible strategies, and we denote respectively, $f_A(\vec{x})$ and $f_B(\vec{x})$ the fitness of both of them. The selection dynamics can be written as

$$\begin{aligned} \dot{x}_A &= x_A[f_A(\vec{x}) - \phi] \\ \dot{x}_B &= x_B[f_B(\vec{x}) - \phi] \end{aligned} \tag{2.35}$$

where $\phi = x_A f_A(\vec{x}) + x_B f_B(\vec{x})$ is the average fitness of the entire population. Obviously, since $x_A + x_B = 1$, we can consider $x \equiv x_A$ and $1 - x \equiv x_B$, and then we can rewrite the previous differential Eq. 2.35 in a simpler way as:

$$\dot{x} = x(1-x)[f_A(x) - f_B(x)] \tag{2.36}$$

It can be easily shown that $x = 0$ is a stable equilibrium if $f_A(0) < f_B(0)$, and conversely, $x = 1$ is a stable equilibrium if $f_A(1) > f_B(1)$. On the other hand, any interior value of $x \in (0, 1)$ is a stable equilibrium x^* if the first derivative of the fitness functions satisfies $f'_a(x^*) < f'_b(x^*)$.

In particular we can calculate the expected fitness of an individual playing A or B respectively, in the well-mixed scenario explained before as:

$$f_A = ax_a + bx_b$$
$$f_B = cx_a + dx_b \tag{2.37}$$

so if we again introduce this expression for the fitness in 2.35 we obtain:

$$\dot{x} = x(1-x)[(a-b-c+d)x + b - d] \tag{2.38}$$

Depending on the relative ordering of the coefficients of the payoff matrix, we can have different situations for the selection dynamics [73, 77, 78]:

(a) *A dominates B*, if $a > c$ and $b > d$. No matter what strategy your opponent uses, it is better always to use A, and selection will lead to a final state where all players are A.
(b) *B dominates A*, if $a < c$ and $b < d$. No matter what strategy your opponent uses, it is better always to use B, and selection will lead to a final state where all players are B.
(c) A and B are *bistable*, if $a > c$ and $b < d$. In this situation, A is the best response for A, and B is the best response for B, so it is better always to try to play the same strategy as your opponent. There is an unstable equilibrium at $x^* = \frac{d-b}{a-b-c+d}$, and depending on the initial fraction of every strategy, the system will converge to all-A (if $x(0) > x^*$) or all-B (if $x(0) < x^*$).
(d) A and B *coexist*, if $a < c$ and $b > d$. In this situation, A is the best response for B, and vice versa, so one should always try to choose the opposite of what the opponent does. Selection will make the system converge to the interior equilibrium $x^* = \frac{d-b}{a-b-c+d}$.
(e) A and B are *neutral*, if $a = c$ and $b = d$. No matter what action you choose, you will always win exactly the same as your opponent, so selection will not modify the initial fraction of every strategy, but this scenario is obviously not very interesting for us.

And some other useful concepts are:

(a) Strategy A is called *risk-dominant* if $a + b > c + d$, and then strategy B has a basin of attraction smaller than $1/2$.
(b) Strategy A is called *pareto-efficient* if $a > d$.
(c) Strategy A is *advantageous* if $a + 2b > c + 2d$, and then strategy B has a basin of attraction smaller than $1/3$.

As a particular example of 2×2 game, we have the Prisoner's Dilemma (see 2.32), that has been widely used to study the phenomenon of cooperation in very different fields, from biology to sociology or economics. It is obvious that defection is the best response, regardless the opponent's (it is in fact, the only Nash equilibrium), despite the fact that, if both cooperate, then they will win more than if both defect.

Thus, both in a Classic Game Theory approach, and in an Evolutionary context using the Replicator Equation we obtain straightforwardly an all-D state, since defectors have higher payoff than cooperators. Cooperation can not survive in a well-mixed situation, it is inevitable. In fact, there are a great deal of examples of this well-mixed or transitory-pairing environments in Nature, which lead to non-cooperative or exploiting situations for the individuals, on the contrary to what usually happens with stable pairing, or even mutualism between different species [59].

Finite Populations

Additionally, one can wonder what happens to the dynamics in the very realistic case of finite populations (notice that we still do not take into account an internal structure). In this case, in order to describe the evolution of a N-sized population, a stochastic theory is needed, and we calculate fixation probabilities for the different possible strategies [73, 79], instead of equilibrium states of the system. The *probability of fixation* of strategy B is the probability of a single mutant B to invade an entire population of A-players.

In order to approach this situation, we can use, among other stochastic processes, the *Moran process* [80], which could be a finite-N analogue to the Replicator Equation. It is a birth-death process that describes the probabilistic dynamics in a finite population of constant size N in which two strategies A and B are competing for dominance. In each time step, a random individual is chosen for reproduction and a random individual is chosen for death; thus ensuring that the population size remains constant. To model selection, one type has to have a higher fitness (considered constant) and is thus more likely to be chosen for reproduction. The same individual can be chosen for death and for reproduction in the same step. It is worth mentioning that in finite populations, even if all different strategies had the same fitness, all but one type will eventually go extinct. This principle is called *neutral drift*. Thus, since coexistence is not possible, there are as many absorbing states as different strategies at the beginning. In a population on size N made up of A individuals, we can calculate [73] the probability of fixation of another strategy B (it is to say, the probability for a single neutral mutant to take over the entire population), and it is given by $1/N$. It means that when dealing with finite populations, just due to random drift, a mutant (with the same fitness as the majority strategy) can invade the system, which is a very different outcome from the infinite-population scenario, where having the same fitness meant coexistence of different strategies. In the same way, the probability of ending up in an all-B state, just due to random drift, when starting with $i \leq N$ individual playing B in a population of A is i/N. On the other hand, if a mutant B has a relative fitness r, with respect to the A players, it can be proven [73] that its probability of fixation is then $\rho = \frac{1-1/r}{1-1/r^N}$. Notice that in this scenario, there is always a non-zero probability that a mutant strategy will invade and take over the whole population, even though it is opposed by selection [81].

2.2.3 Evolution of Cooperation

As we have seen previously, neither within the Classic or the Evolutionary Game approach, can cooperation survive. Nonetheless, there are plenty of examples of real situations where cooperators arise and thrive, so there must be some mechanisms behind it. Over the years, five main ideas [77] have been proposed to help understand this phenomenon: kin selection, direct reciprocity, indirect reciprocity, group selection and network reciprocity.

According to Hamilton [72], natural selection can favor cooperation if the donor and the recipient of an altruistic act are genetic relatives. More precisely, Hamilton's rule establishes that the coefficient of relatedness, r, must exceed the cost-to-benefit ratio of the altruistic act, it is to say: $r > c/b$. This coefficient r is defined as the probability of sharing a gene (it is equal to $1/2$ for siblings, equal to $1/8$ for cousins,...). This theory is called *Kin Selection*, but obviously it can not help understand cooperation among unrelated individuals, or even members of different species.

Trivers proposed the *Direct Reciprocity* mechanism. Let us assume that there are repeated encounters [71] of a the Prisoner's Dilemma Game between the same two individuals, and every time they can choose to be cooperators or defectors. The idea is that if I cooperate in this round of the game, maybe you will cooperate in the next one. When considering the repeated game on a whole population, it can be proven that direct reciprocity leads to the evolution of cooperation only if the probability of another encounter between the same two individuals, w, exceeds the cost-to-benefit ratio of the altruistic act: $w > b/c$.

Let us now consider the following scenario: among a population, two individuals meet once, one of them is in the position of helping the other one (this help is supposed to be less costly for the donor than beneficial for the receiver), and although there is no possibility for direct reciprocation, helping others will establish a good *reputation* which will be rewarded by others. In this way, when deciding how to act, one will take into consideration the consequences for their reputation. Moreover, the next step can be to take into consideration the opponents' reputation, in order to decide whether or not he/she deserves our help, and how it will affect our own. This theory constitutes *Indirect Reciprocity* [82, 83], and when applied to human behavior, it can help understand the origin of moral and social norms.

We can take into account that selection not only acts on the individual level, but also on the group level. A simple model for *Group Selection* is as follows [84]: the population is divided into different groups, and individuals cooperate inside its own group, while defectors do not help anyone. Individuals reproduce proportional to its fitness and the offspring belongs to the same group as the ancestors. When a group reaches certain size, it can split in two, making another group disappear, in order to preserve the total size of the population constant. In a mixed group, a defector reproduces faster than a cooperator, but groups of pure cooperators split faster than those of pure defectors. For the limit of weak selection and considering the case of rare group splitting, it can be obtained that, if n is the maximum group size and

m is the number of groups, then Group Selection allows evolution of cooperation, provided that: $b/c > 1 + (n/m)$, where b/c is the cost-to-benefit ratio.

Finally, one can realize that the Evolutionary approach for the PD game always leads to all-D situations, but it considers a well-mixed scenario, it is to say, at any given time, every individual has equal probabilities to interact with everyone else. Nonetheless we know that this is a very unrealistic assumption, since groups and societies have usually some kind of internal structure. In other words, there is a well defined pattern of interactions among individuals, so every one of them has a fixed number of neighbors. It has been shown that spatial structure affects greatly the outcome of an evolutionary dynamics, allowing cooperators to survive in many situations. Specifically, cooperators form network clusters, where they help each other. The analytical treatment of this problem is hard, and many times, even impossible, but it has been found that this *Network Reciprocity* can favor cooperation if $b/c > k$, where k stands for the average number of connections of the individuals in the population.

Prisoner's Dilemma Game on Structured Populations

According to what we have seen previously, one of the mechanisms that helps promote cooperation is Network Reciprocity, and it happens to be also the one we will be interested during this thesis. Thus, the natural next step for us in order to build more realistic models of social or biological interactions, is to consider some sort of underlying structure, in account for the particular pattern of relationships between individuals. The first attempts to model such social structure for the Prisoner's Dilemma game considered the individuals placed in a regular lattice [85–91]. Those studies found that spatial structure affects greatly the outcome of such dynamics. Specifically, by making the agents play just with a small number of fixed neighbors, we can make cooperation and defection coexist, or even enhance cooperation. In fact, when dealing with games in spatial structure populations, the equilibria among strategies are no longer necessarily characterized by their having equal average payoff. Instead, the asymptotic equilibrium properties are now determined by 'local relative payoffs', and not by global averages [88]. It was also found for the PD in lattices, that under certain symmetrical initial conditions for the distribution of strategies, certain values of the temptation to defect b, and as long as we use deterministic updating rules, kaleidoscopic carpet-like chaotically-changing spatial patterns arise [86, 87]. Moreover, it has been found that there is a critical phase transition in the Prisoner's Dilemma game in lattices that falls into the same universality class than directed percolation [91].

Some effort was put also on the analytical study of how different kind of structures can favor fixation of the strategies or, on the contrary, favor neutral drift, explicitly calculating to that end the corresponding probabilities of fixation of the strategies on some networks with very particular topologies, such as stars, paths, downstreams, upstreams or funnels [73, 76, 92]. Moreover, striking results in terms of survival of cooperation were found for random and SF networks, but for such general structures, no explicit calculations can be performed, so one needs to rely totally on simulations. In this area, a great deal of effort has been put too, and as a very general remark, it can be said that the complex topologies behind the interactions among a given population affect the outcome of any process [29, 30, 36, 40–42, 93] not only games [44, 76, 86, 92] to a large extent. Specifically, as we will see with some detail in Chap. 3, when it comes to the Prisoner's Dilemma game on complex networks, a large number of studies [66, 94–98] have pointed out that cooperation benefits from heterogeneity. It is to say, it has much better chances to survive in scale-free than in random topologies, for the same given values of the parameters of the game.

References

1. S. Boccaletti, V. Latora, Y. Moreno, M. Chavez, and D. U. Hwang, Phys. Rep. **424**, 175 (2006).
2. M. E. J. Newman, SIAM Review **45**, 167 (2003).
3. J. Dunne, R. Williams, and N. Martinez, Marine Ecological Press Series **273**, 291 (2004).
4. M. E. J. Newman, Proc. Natl. Acad. Sci. USA **98**, 404 (2001).
5. A. Arenas, L. Danon, A. Díaz-Guilera, P. Gleiser, and R. Guimerá European Physical Journal B **38**(**2**), 373 (2004).
6. R. Guimerá and M. Sales-Pardo, Proc. Natl. Acad. Sci. USA **106**, 22073 (2009).
7. S. Gómez, P. Jensen, and A. Arenas, Phys. Rev. E **80**, 016114 (2009).
8. M. Newman and M. Girvan, Phys. Rev. E **69**, 026113 (2004).
9. L. Danon, A. Díaz-Guilera, J. Duch, and A. Arenas, J. Stat. Mech. p. P09008 (2005).
10. A. Barabási and Z. Oltvai Nature Reviews, Genetics **5**, 101 (2004).
11. J. Duch, and A. Arenas Phys. Rev. E **72**, 027104 (2005).
12. L. Danon, J. Duch, A. Arenas, and A. Díaz-Guilera, World Scientific p. 93 (2007).
13. M. Sales-Pardo, R. Guimerà, A. Moreira, and L. Amaral Proc. Natl. Acad. Sci. USA **104**, 15224 (2007).
14. P. Erdős, and A. Renyi, Publicationes Mathematicae Debrecen **6**, 290 (1959).
15. D. J. Watts and S. H. Strogatz, Nature **393**, 440 (1998).
16. S. Milgram, Psycol. Today **2**, 60 (1967).
17. J. Guare, *Six degrees of separation: a play*. (Vintage Books, New York, 1990).
18. J. Travers and S. Milgram, Sociometry **32**, 425 (1969).
19. L. Amaral, A.Scala, M. Barthélémy, and H. Stanley, Proc. Natl. Acad. Sci. USA **97**, 11149 (2000).
20. M. Barthélémy and L. Amaral, Phys. Rev. Lett. **82**, 3180 (1999).
21. H. Simon, Biometrika 42, 425 (1955).
22. R. Merton, Science **159**, 56 (1968).
23. R. Albert, and A. L. Barabási, Rev. Mod. Phys. **74**, 47 (2002).
24. S. N. Dorogovtsev, J. F. Mendes, and A. N. Samukhin, Phys. Rev. Lett. **85**, 4633 (2000).
25. S. N. Dorogovtsev, and J. F. F. Mendes, *Evolution of networks. From biological nets to the Internet and the WWW*. (Oxford University Press, Oxford, UK, 2003).
26. P. Holme, and B. Kim, Phys. Rev. E **65**, 026107 (2002).
27. G. Caldarelli, A. Capocci, P. D. L. Rios, and M. A. M. noz, Phys. Rev. Lett. **89**, 258702 (2002).
28. J. Gómez-Gardeñes and Y. Moreno, Phys. Rev. E **73**, 056124 (2006).
29. R. Pastor-Satorras, and A. Vespignani, Phys. Rev. Lett. **86**, 3200 (2001a).
30. R. Pastor-Satorras and A. Vespignani, Phys. Rev. E **63**, 066117 (2001b).
31. Y. Moreno, R. Pastor-Satorras, and A. Vespignani, European Physical Journal B **26**, 521 (2002).
32. R. Pastor-Satorras, and A. Vespignani, Phys. Rev. E **65**, 036104 (2002).

33. M. Boguñá, R. Pastor-Satorras, and A. Vespignani, Phys. Rev. Lett. **90**, 028701 (2003).
34. R. May, and A. Lloyd, Phys. Rev. E **64**, 066112 (2001).
35. M. Newman, Phys. Rev. E **66**, 016128 (2002).
36. J. Gómez-Gardeñes, Y. Moreno, and A. Arenas, Phys. Rev. Lett. **98**, 034101 (2007a).
37. L. Donetti, P. I. Hurtado, and M. A. Muñoz, Phys. Rev. Lett. **95**, 188701 (2005).
38. J. Gómez-Gardeñes, and Y. Moreno, Int. J. Bifurcation Chaos **17**, 2501 (2007).
39. J. Gómez-Gardeñes, Y. Moreno, and A. Arenas, Phys. Rev. E **75**, 066106 (2007b).
40. A. Arenas, A. Díaz-Guilera, J. Kurths, Y. Moreno, and C. Zhou, Phys. Rep. **469**, 93 (2008).
41. R. Cohen, K. Erez, D. ben-Avraham, and S. Havlin, Phys. Rev. Lett. **86**, 3682 (2001).
42. R. Cohen, K. Erez, D. ben-Avraham, and S. Havlin, Phys. Rev. Lett. **85**, 4626 (2000).
43. C. Castellano, S. Fortunato, and V. Loreto, Rev. Mod. Phys. **81**, 591 (2009).
44. G. Szabó and G. Fáth, Phys. Rep. **446**, 97 (2007).
45. V. Colizza, A. Barrat, M. Barthelemy, A. Valleron, and A. Vespignani, PLos Medicine **4**, e13 (2007).
46. R. Axelrod, J. Conflict Res. **41**, 203 (1997b).
47. K. Klemm, V. Equíluz, R. Toral, and M. San Miguel, Phys. Rev. E **67**, 026120 (2003a).
48. D. Vilone, A. Vespignani, and C. Castellano, European Physical Journal B **30**, 399 (2002).
49. K. Klemm, V. Eguiluz, R. Toral, and M. San Miguel, Physica A **327**, 1 (2003b).
50. J. González-Avella, V. Eguíluz, M. Cosenza, K. Klemm, J. Herrera, and M. San Miguel, Phys. Rev. E **73**, 046119 (2006).
51. C. Castellano, M. Marsili, and A. Vespignani, Phys. Rev. Lett. **85**, 3536 (2000).
52. B. Guerra, J. Poncela, J. Gómez-Gardeñes, V. Latora, and Y. Moreno, Phys. Rev. E **81**, 056105 (2010).
53. K. Klemm, V. Eguiluz, R. Toral, and M. San Miguel, J. Economic Dynamics and Control **29**, 321 (2005).
54. K. Klemm, V. Eguíluz, R. Toral, and M. San Miguel, Phys. Rev. E **67**, 045101(R) (2003c).
55. J. Gonzalez-Avella, M. Cosenza, V. Eguíluz, and M. San Miguel, New J. Phys. **12**, 013010 (2010).
56. A. Grabowski, and R. A. Kosinski, Phys. Rev. E **73**, 016135 (2006).
57. C. Gracia-Lázaro, L. Lafuerza, L. Floria, and Y. Moreno, Phys. Rev. E **80**, 046123 (2009).
58. T. Schelling, The American Economic Review **59**, 488 (1969).
59. R. Axelrod, and W. Hamilton, Science **211**, 1390 (1981).
60. J. von Neumann, and O. Morgenstern, *Theory of Games and Economic Behavior*. (Princeton University Press, Princeton, NJ, 1944).
61. J. Maynard Smith, and G. Price, Nature **246**, 15 (1973).
62. H. Gintis, *Game theory evolving*. (Princeton University Press, Princeton, NJ, 2000).
63. J. Hofbauer, and K. Sigmund, *Evolutionary games and population dynamics*. (Cambridge University Press, Cambridge, UK, 1998).
64. J. Nash, Proc. Natl. Acad. Sci. USA **36**, 48 (1950).
65. T. Killingback, and M. Doebeli, Proc. R. Soc. Lond. **263**, 1135 (1996).
66. C. P. Roca, J. A. Cuesta, and A. Sánchez, Phys. Rev. E **80**, 046106 (2009).
67. M. Tomassini, L. Luthi, and M. Giacobini, Phys. Rev. E **73**, 106 (2006).
68. C. Hauert, and M. Doebeli, Nature **428**, 643 (2004).
69. L. E. Blume, Games and Economic Behavior **5**, 387 (1993).
70. G. Ellison, Econometrica **61**, 1047 (1993).
71. R. Axelrod, *The Evolution of Cooperation*. (Basic Books, New York, 1984).
72. W. Hamilton, J. Theor. Biol. **7**, 1 (1964).
73. M. Nowak, *Evolutionary dynamics: exploring the equations of life*. (Harvard University Press, Cambridge, MA, 2006b).
74. J. Hofbauer, P. Schuster, and K. Sigmund, J. Theor. Biol. **81**, 609 (1979).
75. P. Taylor, and L. Jonker, Math. Biosci. **40**, 145 (1978).
76. H. Ohtsuki, and M. A. Nowak, J. Theor. Biol. **243**, 86 (2006).
77. M. Nowak, Science **314**, 1560 (2006a).
78. M. A. Nowak, and K. Sigmund, Science **303**, 793 (2004).

79. A. Traulsen, M. Nowak, and J. Pacheco, Phys. Rev. E **74**, 011909 (2006).
80. P. Moran, Mathematical Proceedings of the Cambridge Philosophical Society **54**, 60 (1958).
81. M. Nowak, A. Sasaki, C. Taylor, and D. Fudenberg, Nature **428**, 646 (2004).
82. M. Nowak, and K. Sigmund, Nature **393**, 573 (1998).
83. M. Nowak, and K. Sigmund, Nature **437**, 1291 (2005).
84. A. Traulsen, and M. A. Nowak, Proc. Natl. Acad. Sci. USA **103**, 10952 (2006).
85. M. Nowak and K. Sigmund, Games on Grids, *in: The Geometry of Ecological Interactions.* (Cambridge University Press, Cambridge, UK, 2000).
86. M. A. Nowak and R. M. May, Nature **359**, 826 (1992).
87. M. Nowak, S. Bonhoeffer, and R. May, Int. J. Bifurcation Chaos **4**, 33 (1994).
88. M. Nowak, S. Bonhoeffer, and R. May, Proc. Natl. Acad. Sci. USA **91**, 4877 (1994).
89. G. Szabó and C. Tőke, Phys. Rev. E **58**, 69 (1998).
90. M. Nakamaru, H. Matsuda, and Y. Iwasa, J. Theor. Biol. **184**, 65 (1997).
91. C. Hauert and G. Szabó, Am. J. Phys. **73**, 405 (2005).
92. E. Lieberman, C. Hauert, and M. A. Nowak, Nature **433**, 312 (2005).
93. D. Callaway, M. Newman, S. Strogatz, and D. Watts, Phys. Rev. Lett. **85**, 5468 (2000).
94. G. Abramson, and M. Kuperman, Phys. Rev. E **63**, 030901(R) (2001).
95. F. C. Santos, and J. M. Pacheco, Phys. Rev. Lett. **95**, 098104 (2005).
96. F. C. Santos, F. J. Rodrigues, and J. M. Pacheco, Proc. Biol. Sci. **273**, 51 (2006a).
97. H. Ohtsuki, E. L. C. Hauert, and M. A. Nowak, Nature **441**, 502 (2006).
98. F. C. Santos, and J. M. Pacheco, J. Evol. Biol. **19**, 726 (2006).

Part I
Evolutionary Dynamics on Static Complex Networks

Presentation of Part I

In this first part of the Thesis, we want to focus on the effect that the structure of interactions among the constituents of a given complex system has on the evolutionary dynamics that takes place on top of it. On the one hand, the individuals of the system form a complex network [1–6, 7, 8], that could represent a very simple version of a society or a social organization [9, 10] of humans or other species. On the other hand, the kind of dynamics we will be taking into consideration is dictated by Evolutionary Game Theory [11–14]. We will focus on the situation in which nodes represent individuals engaged with their neighbors in a certain (2×2) game, using a certain strategy that can be updated after every round of the game, depending on the outcome of it. In other words, the outcome of the game, meaning the accumulated payoff every node gets in a single round, will affect the probability of maintaining or changing its strategy for the next round of the game. This can also be interpreted in terms of evolutionary fitness and reproduction of the individuals: instead of considering individuals of a population that update their strategies for the next round of the game, one can also think of the benefits of an individual in terms of its reproductive success or fitness, meaning the probability of its offspring to be present in the system in the next generation, using its very same strategy [15]. In this way, we are not specially interested in the evolution of a particular node, but in the entire population as a whole, and to this end, we will measure the proportion of the different strategies that are present in the stationary state of the dynamics, as well as its microscopical organization within the network.

Specifically, in Chap. 3 we will study in detail the outcome of the (weak) Prisoner's Dilemma game [16–18, 15, 19, 11, 20–23, 12] on top of complex networks [24–26, 27–29], comparing the results obtained mainly for two kind of topologies: ER [30] and BA [1] networks. We will also consider the same dynamics on top of some other systems with intermediate degree of heterogeneity. On the one hand, in order to confirm and understand the well-established fact that cooperation is enhanced by the heterogeneity of the underlying graph [31, 32–42], we will look into the microscopic organization of cooperation in the stationary state, studying the formation of clusters for both strategies. We will find that this organization is quite different depending on the kind of network we are dealing with. We will also analyze the level of cooperation for every connectivity class, for the case of heterogeneous graphs, finding there a plausible explanation for the high levels of cooperation these particular structures can sustain. On the other hand, we will show the asymptotic existence of pure strategists and fluctuating individuals. Moreover, we will prove it by using a simplified but general enough case of a graph (Dipolar Model), where some analytical calculations can be performed.

In Chap. 4 we will expand all these studies not only to the general Prisoner's Dilemma, but also to the Hawks and Doves game [16, 35, 39, 42–44, 45–49], comparing the results with the ones found previously for the weak Prisoner's Dilemma. Analogously to Chap. 3, we will study the stationary state of the system, the level of cooperation it can achieve, the microscopic organization of the different strategies and the formation of strategic clusters. All of it will be considered depending on the underlying topology, remarking the differences found not only between homogeneous and heterogeneous graphs, but also between the Prisoner's Dilemma game and the Hawks and Doves game.

In Chap. 5, we want to address the issue of cooperation in random scalefree networks, comparing the level of cooperation obtained in such correlationfree heterogeneous topologies with those corresponding to the BA networks, in order to confirm the role that the correlations among nodes [6, 42, 50, 7] may play on the sustenance of the level of cooperation in the system [32, 34]. On the other hand, we will propose a degree-based mean-field approach to try to explain the outcome of the Prisoner's Dilemma dynamics on top of random SF networks. We will make further a compartmentalization of the fraction of cooperators and defectors into different connectivity classes, to formulate a set of differential equations for the time evolution of the fraction of cooperators in each degree class. The idea behind this approach is inspired by several works focused on the study of disease spreading on an heterogeneous population, using a similar theoretical framework [51–53]. Thus, we will compare the analytical results with the conventional numerical simulations performed on top of such random SF graphs. We will analyze this in a general case, where we will find that the theoretical approximation and the numerical simulations do not agree. However, we will also explore some particular initial conditions, where cooperators are not placed initially at random, but occupying the largest degrees of connectivity (targeted cooperation). In this latter case we will be able to reproduce (up to an extent) the results from a simulation on top of random SF graphs using these analytical calculations.

Finally, in Chap. 6 we will propose a more realistic scenario for a population with a complex pattern of connections engaged in an evolutionary dynamics such as the Prisoner's Dilemma. The set of individuals will form a network of social contacts, namely a scale-free graph, and will play the game with their neighbors as usual. Nonetheless, we will consider a restriction in the number of interactions a node can sustain in every round of the game. To our knowledge, there are not any works addressing this particular issue, apart from [33], where a cutoff is imposed to the degree distribution of a SF network. However, we will not proceed by altering the degree distribution of the underlying topology. Instead, we will force the nodes to choose randomly a different selection among its topological neighbors for every round of the game. In this way, we want to acknowledge the fact that the amount of energy and time an individual can spend interacting with its neighbors is finite, so the number of acquaintances it interacts with per unit of time should not be given just by its topological connectivity, but it also should be subject to some kind of practical limitations. We will find some striking results that point out that in a situation with some degree of restriction in the number of interactions allowed per node and per round of the game, cooperation can be enhanced even more than in an unrestricted scale-free scenario, when participation costs are also introduced in the formulation of the evolutionary game.

References

1. A. Barabási and R. Albert, Science **286**, 509 (1999)
2. S. H. Strogatz, Nature **410**, 268 (2001)
3. D. J. Watts and S. H. Strogatz, Nature **393**, 440 (1998)
4. M. Newman, SIAM Review **45**, 167 (2003).
5. S. Boccaletti, V. Latora, Y. Moreno, M. Chavez, and D. Hwang, Phys. Rep. **424**, 175 (2006)
6. R. Albert and A. L. Barabási, Rev. Mod. Phys. 74, 47 (2002)
7. S. N. Dorogovtsev and J. F. F. Mendes, *Evolution of networks. From biological nets to the Internet and the WWW*. (Oxford University Press, Oxford, UK, 2003)
8. S. Bornholdt and H. G. Shuster, *Handbook of graphs and networks*. (Wile-VCH, Germany, 2003)
9. A. Arenas, L. Danon, A. Díaz-Guilera, P. Gleiser, and R. Guimerá, EuropeanPhysical Journal B **38(2)**, 373 (2004)
10. M. Newman, Proc. Natl. Acad. Sci. USA **98**, 404 (2001)
11. J. Hofbauer and K. Sigmund, *Evolutionary games and population dynamics*. (Cambridge University Press, Cambridge, UK, 1998)
12. M. Nowak, *Evolutionary dynamics: exploring the equations of life*. (Harvard University Press., Cambridge, MA, 2006)
13. J. Maynard Smith and G. Price, Nature **246**, 15 (1973)
14. H. Gintis, *Game theory evolving*. (Princeton University Press, Princeton,NJ, 2000)
15. R. Axelrod and W. Hamilton, Science **211**, 1390 (1981)
16. M. Nowak and K. Sigmund, Nature **437**, 1291 (2005)
17. R. Axelrod, *The Evolution of Cooperation*. (Basic Books, New York, 1984)
18. W. Hamilton, J. Theor. Biol. **7**, 1 (1964)
19. M. Nowak, Science **314**, 1560 (2006)
20. J. Hofbauer and K. Sigmund, Bull. Am. Math. Soc. **40**, 479 (2003)
21. M. Nowak and K. Sigmund, Nature **355**, 250 (1992)

22. M. Nowak and K. Sigmund, Acta Applicandae Math **20**, 247 (1990)
23. R. Axelrod, *The complexity of cooperation: agent-based models of competitionand collaboration.* (Princeton University Press., Princeton, NJ, 1997)
24. M. A. Nowak and R. M. May, Nature **359**, 826 (1992)
25. M. Nowak, S. Bonhoeffer, and R. May, Int. J. Bifurcation Chaos **4**, 33(1994)
26. M. Nowak, S. Bonhoeffer, and R. May, Proc. Natl. Acad. Sci. USA **91**, 4877 (1994)
27. M. Nowak and K. Sigmund, *Games on Grids, in: The Geometry of Ecological Interactions.* (Cambridge University Press, Cambridge, UK, 2000)
28. G. Szabó and C. Tőke, Phys. Rev. E **58**, 69 (1998)
29. M. Nakamaru, H. Matsuda, and Y. Iwasa, J. Theor. Biol. **184**, 65 (1997)
30. P. Erdős and A. Reńyi, Publicationes Mathematicae Debrecen **6**, 290 (1959)
31. G. Szabó and G. Fáth, Phys. Rep. **446**, 97 (2007)
32. F. C. Santos and J. M. Pacheco, Phys. Rev. Lett. **95**, 098104 (2005)
33. F. C. Santos, F. J. Rodrigues, and J. M. Pacheco, Proc. Biol. Sci. **273**, 51 (2006)
34. F. C. Santos and J. M. Pacheco, J. Evol. Biol. **19**, 726 (2006)
35. F. C. Santos, J. M. Pacheco, and T. Lenaerts, Proc. Natl. Acad. Sci. USA **103**, 3490 (2006)
35. H. Ohtsuki, E. L. C. Hauert, and M. A. Nowak, Nature **441**, 502 (2006)
36. G. Abramson and M. Kuperman, Phys. Rev. E **63**, 030901(R) (2001)
37. V. M. Eguíluz, M. G. Zimmermann, C. J. Cela-Conde, and M. San Miguel, American Journal of Sociology **110**, 977 (2005)
38. T. Killingback and M. Doebeli, Proc. R. Soc. Lond. **263**, 1135 (1996)
39. A. Szolnoki, M. Perc, and Z. Danku, Physica A **387**, 2075 (2008)
40. J. Vukov and G. S. A. Szolnoki, Phys. Rev. E **77**, 026109 (2008)
41. J. Gómez-Gardeñes, M. Campillo, L. M. Floría, and Y. Moreno, Phys. Rev. Lett. **98**, 108103 (2007)
42. C. P. Roca, J. A. Cuesta, and A. Sánchez, Phys. Rev. E **80**, 046106 (2009)
43. M. Tomassini, L. Luthi, and M. Giacobini, Phys. Rev. E **73**, 106 (2006)
44. C. Hauert and M. Doebeli, Nature **428**, 643 (2004)
45. M. Sysi-Aho, J. Saramaki, J. Kertész, and K. Kaski, European Physical Journal B **44**, 129 (2005)
46. L. Zhong, D. Zheng, B. Zheng, C. Xu, and P. Hui, Europhys. Lett. **76**,724 (2006)
47. A. Kun, G. Boza, and I. Scheuring, Behav. Ecol. **17**, 633 (2006)
48. F. C. Santos, J. M. Pacheco, and T. Lenaerts, PLos Comput. Biol. **2**(**10**), e140 (2006)
49. J. Maynard Smith, *Evolution and the Theory of Games.* (Cambridge University Press, Cambridge, UK, 1982)
50. S. N. Dorogovtsev, J. F. Mendes, and A. N. Samukhin, Phys. Rev. Lett. **85**, 4633 (2000)
51. R. Pastor-Satorras and A. Vespignani., Phys. Rev. Lett. **86**, 3200 (2001)
52. R. Pastor-Satorras and A. Vespignani., Phys. Rev. E **63**, 066117 (2001)
53. Y. Moreno, R. Pastor-Satorras, and A. Vespignani., European Physical Journal B **26**, 521 (2002)

Chapter 3
The Prisoner's Dilemma on Static Complex Networks

The PD game has been frequently used [1–5] when trying to model the emergence of cooperative behavior in a social or biological system. The questions of why and how cooperation arises and survives in an environment where it is clearly more expensive for the individual than defection in the short term have been subject of intense research for quite some time, and the PD turned out to be a very useful tool for this aim. One of the aspects that have been pointed out as responsible for the survival of cooperation is, among others, the so-called network reciprocity [6]. Several studies have shown that cooperation can be greatly promoted by placing the individuals of a population on the nodes of a network of contacts, instead of letting them interact in a well-mixed situation, where no asymptotic cooperation exists. First, some effort was put on studying the PD on regular lattices, finding that, as long as the connectivity of the nodes was not to high, cooperation actually got a chance at survival (however, when the number of neighbors increases, the situation resemblances more and more an all-to-all scenario, and cooperation dies out again). Next, PD was studied in complex topologies [7–18], in an attempt to model more accurately the pattern of connections of a real system, and this is precisely the problem we will consider in this chapter of the thesis.

In this way, we want to address the dependence of the PD dynamics on the topology of the underlaying structure. As we have already advanced, we are interested in characterizing the final equilibrium state that the system achieves when implementing the dynamics of such structures, namely random and SF graphs, paying special attention not only to the asymptotic level of cooperation, but more important, to the microscopic organization of the strategies. This is actually, as we will see in detail, the key point of the differences found between both topologies when it comes to the asymptotic level of cooperation. We will also take care of other aspects of the dynamics, such as the dependence of the final level of cooperators in the system with the initial fraction of them, or the distribution of strategies according to the different classes of connectivity for SF networks.

J. Poncela Casasnovas, *Evolutionary Games in Complex Topologies*, Springer Theses,
DOI: 10.1007/978-3-642-30117-9_3, © Springer-Verlag Berlin Heidelberg 2012

3.1 The Model

The Prisoner's Dilemma is a two-player game defined in its more general form by the payoff matrix (see Sect. 2.2):

$$\begin{array}{c c} & \begin{array}{cc} C & D \end{array} \\ \begin{array}{c} C \\ D \end{array} & \begin{pmatrix} R & S \\ T & P \end{pmatrix} \end{array} \tag{3.1}$$

where the element a_{ij} is the payoff received by an i-strategist when playing against a j-strategist, with $i = 1$ meaning cooperator (C), and $i = 2$ defector (D). Thus, both receive R (Reward) under mutual cooperation and P (Punishment) under mutual defection, while a cooperator receives S (Sucker's Payoff) when confronted to a defector, which in turn receives T (Temptation to defect). The payoff ordering is given by $T > R > P > S$. Under these conditions, defection is the best response regardless the opponent's strategy. Indeed, in a well-mixed population of N replicators, i.e. where every individual interacts with everyone else, the defection strategy is unbeatable and reaches fixation. However, if individuals only interact with its k_i neighbors, as dictated by the underlying network of contacts, it has been proven the asymptotic survival of cooperation for $T \geq R$ on different types of complex topologies [7–18].

Following several studies [7, 8, 18–20], we set the PD payoffs to $R = 1$ (so the reward for cooperating fixes the payoff scale), $T = b > 1$, $P = 0$ (no benefit under mutual defection), and $P - S = \epsilon \to 0^+$. This last choice places us in the very frontier of PD game, or the 'weak' Prisoner's Dilemma. It has the effect of not favoring any strategy when playing against defectors (while being advantageous to play defection against cooperators). Small positive values of the parameter $\epsilon \ll 1$ leads to no qualitative differences in the results [19], so the limit $\epsilon \to 0^+$ is agreed to be continuous.

The dynamic rule is specified as follows: each time step is thought of as one generation of the discrete evolutionary time, where every node i of the system plays with its nearest k_i neighbors (given by the underlying network) and accumulates the payoffs obtained during the round, say P_i. As Evolutionary Game Theory approach dictates, the benefit an agent gets from the game should be interpreted as its fitness in the Darwinian sense of reproductive success [2, 21]. Specifically, we consider that individuals are then allowed to synchronously change their strategies by comparing the payoffs they accumulated in the previous generation with that of a neighbor j chosen at random. If $P_i > P_j$, player i keeps the same strategy for the next time step, when it will play again with all of its neighbors. On the contrary, whenever $P_j > P_i$, i adopts the strategy of j with probability

$$\Pi_{i \to j} = \frac{P_j - P_i}{\max\{k_i, k_j\} b} \tag{3.2}$$

Following previous studies, we called this updating rule *Replicator-like* [4, 5, 8, 9, 22, 23], because it is obviously similar to the Replicator Equation (see Sect. 2.2.2): the probability of changing strategy is proportional to the difference of payoffs of the nodes involved, and it is normalized by the maximum payoff a node can get, i.e., b times its connectivity. Note also that this dynamic rule, though stochastic, does not allow the adoption of irrational strategy, i.e., $\Pi_{i \to j} = 0$ whenever $P_j \leq P_i$.

Regarding the synchrony of the strategy updating of the individuals in the population, it is worth mentioning here that we have not found significant differences when comparing to asynchronous updating (also known as sequential updating or *continuous time*), and thus in good agreement with previous findings for this particular PD game and Replicator-like rule [24], in spite of the fact that one can always argue that synchronous or asynchronous updating more accurate in order to describe different biological or social scenarios, respectively [20].

Let's now specify precisely the family of networks on top of which the evolutionary PD game is evolving. Strategists are located on the vertexes of a fixed graph of average connectivity $\langle k \rangle = 4$. The heterogeneity of the networks is controlled by tuning a single parameter α, according to the recipe introduced by Gardeñes-Moreno (GM) in [25]. As we explained in detail in Sect. 2.1.3, the GM model creates a network by combining the mechanisms of preferential attachment with probability α and uniform random linking with probability $1 - \alpha$. Thus, in this model, when $\alpha = 0$ the generated networks are of the ER [26] class of random graphs, and when $\alpha = 1$ they are of the BA [27] scale-free networks class. On the other hand, networks with an intermediate degree of heterogeneity can be built with $0 < \alpha < 1$. We will study the dynamics on top of such networks with intermediate heterogeneity at the end of this chapter (see Sect. 3.9), but for now, we will focus just on the extreme cases $\alpha = 0$ and $\alpha = 1$. It is also worth stressing that the different topologies we will compare during this chapter have always the same number of nodes, N, and average connectivity $\langle k \rangle$.

3.2 Dynamic Equilibrium

The initial strategy of each one of the N nodes is randomly set, with a probability of being a cooperator $\rho_0 = 0.5$ (note that ρ_0 is also the initial fraction of cooperation on the system), and then the dynamics starts. We let the system evolve for 5×10^3 time steps or generations, after which we check whether the equilibrium has been reached. To do so, we observe the time evolution of the fraction of cooperators, $c(t)$, during a time window of 10^3 generations. If the slope of $c(t)$ is smaller than 10^{-2}, then we consider the equilibrium has been reached. Otherwise, we let the system evolve 5×10^3 more generations, after which, we will evaluate the equilibrium condition again.

We show several examples of temporal evolution of the system in Fig. 3.1. The behavior during the transient time of the fraction of cooperators in the system can be understood as follows: as we have said, the system starts with a fraction of ρ_0

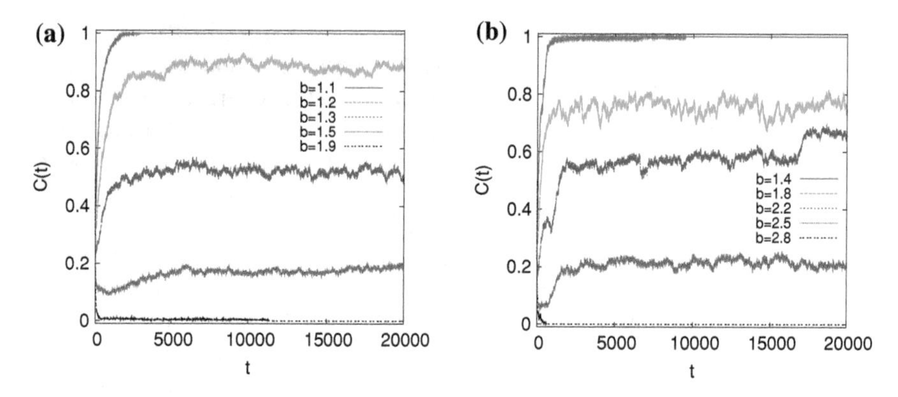

Fig. 3.1 Several examples of the temporal evolution of the level of cooperation in the system for ER (**a**) and SF (**b**) networks as a function of b. The size of the networks is $N = 4 \times 10^3$ nodes and average connectivity $\langle k \rangle = 4$

cooperators, randomly distributed on the network. The defectors take advantage of this initial situation, getting very high payoff exploiting its cooperator neighbors, and forcing other nodes to imitate them. Therefore, the level of cooperation drops initially. However, after a few more time steps, the defectors are surrounded by more defectors, and they can not get benefits anymore, while cooperators start clustering themselves, and providing payoff from one another. Thus, cooperators self-organize and hold a non-negligible level of cooperation on the network. As it can be seen in Fig. 3.1, the macroscopic behavior of the system towards its dynamical equilibrium is qualitatively very similar, regardless the underlying topology. Nevertheless, as we will explain later in detail in Sect. 3.7, the microscopic organization of cooperators and defectors when the equilibrium has been reached is very different depending on the network, and it is specially non-trivial for BA networks.

From any initial condition for the whole system $\{s_i(t = 0)\}$ (with $i = 1, \ldots, N$, and where $s_i = 1$ if node i is an instantaneous cooperator and $s_i = 0$ if it is a defector in that step), and after many generations, the instantaneous fraction of cooperators, given by

$$c(t) = N^{-1} \sum_{i=1}^{N} s_i(t) \tag{3.3}$$

in the stochastic trajectory, $\{s_i(t)\}$, fluctuates around a well-defined mean value $\langle c \rangle$. In turn, this average value of cooperation can be defined as follows:

$$\langle c \rangle = \frac{1}{T} \sum_{\tau=t_0}^{t_0+T} c(\tau), \tag{3.4}$$

where t_0 is the transient period, and T is the period of time during which we observe the system, once it has reached the equilibrium. Thus, this average level of cooperation depends only on the value of the parameter b, and the initial fraction of cooperators ρ_0 (and also on the topology of the system, as we will see). The average level of cooperation $\langle c \rangle$ is computed as the average of $\langle c \rangle$ over 10^3 independent realizations with different initial conditions (different random distributions of a fixed value for the fraction ρ_0 of cooperators, as well as network realizations).

It is worth mentioning that the *time scale* of microscopic invasion processes, it is to say, the pace of the updating rule for any given node, is controlled by

$$\beta^{-1} = \max\{k_i, k_j\}b, \tag{3.5}$$

which is essentially determined by the highest connectivity of the pair of nodes we take under consideration. This makes that the very high payoff of a hub (due to its very high k) to be balanced by $\beta \propto k^{-1}$ [8, 9, 18], with the side effect that the invasion processes from and to hubs are slowed down, if hub's (and neighbor's) payoff is much smaller than its connectivity k. On the other hand, the transient time t_0 should be greater than characteristic fixation times for the nodes, if one is interested in measuring observable quantities associated to the dynamical equilibrium.

3.3 Pure Strategists and Fluctuating Individuals

After the transient time t_0 has passed, we establish a 10^4 time step window during which we measure the relevant magnitudes of the system. This procedure allows us to scrutinize in depth the microscopic temporal evolution of cooperation as well as to characterize how its local patterns are formed. We note that individual's strategies asymptotically (i.e. $t > t_0$) follow three different behaviors. Let $P(x,t)$ be the probability that a node adopts the strategy x at any time $t > t_0$. We say that an element i of the population is *pure cooperator* (PC) if $P(s_i = 1, t) = 1$, i.e., it plays as cooperator in all generations after the transient time. Conversely, *pure defectors* (PD) are those individuals for which $P(s_i = 0, t) = 1$. And there is a third set, constituted by *fluctuating nodes* (F) which are those that are neither pure cooperators nor pure defectors, so they spend alternatively some time as cooperators and some time as defectors. This set is what was first called 'unsatisfied elements' by Abramson and Kuperman in [11].

From now on, we denote by $\rho_C = \langle \mu(PC) \rangle$ the measure (relative size) of the set of pure cooperators (averaged over initial conditions and network realizations), by $\rho_D = \langle \mu(PD) \rangle$ that of the set of pure defectors, and by $\rho_F = \langle \mu(F) \rangle$ that of the set of fluctuating strategists. At any given time during the simulation, the relation between the fractions $\rho_C + \rho_D + \rho_F = 1$ must be fulfilled by the system, obviously.

On the other hand, the macroscopic average level of cooperation $\langle c \rangle$ can be written as:

$$\langle c \rangle = \rho_C + \rho_F \langle T_C \rangle \tag{3.6}$$

where $\langle T_C \rangle$ is the average proportion of time spent by the fluctuating subpopulation as cooperators (see Sect. 3.6 for further details).

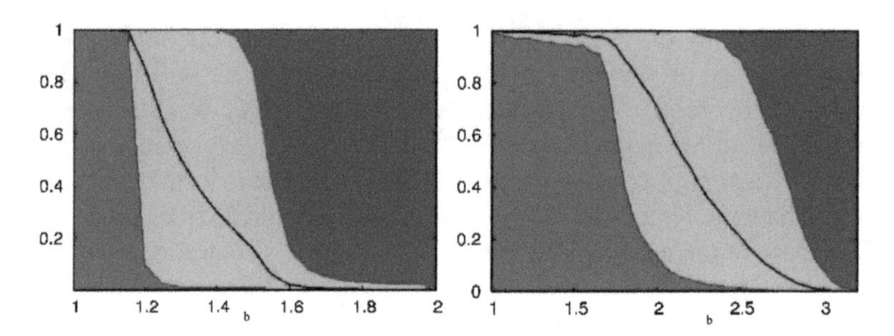

Fig. 3.2 Fraction ρ_C of pure cooperators (*Red Area*), fraction ρ_D of pure defectors (*Blue Area*), fraction ρ_F of fluctuating nodes (*Green Area*) and the average level of cooperation $\langle c \rangle$ in the system (*Solid black line*) as a function of b for ER networks (*Left*) and BA networks (*Rigth*). The size of the networks is $N = 4 \cdot 10^3$ nodes and average connectivity $\langle k \rangle = 4$

In the Fig. 3.2 we show the fraction of pure strategists and fluctuating individuals, and the average level of cooperation as a function of b, for BA and ER networks. As one could expect, both the average level of cooperation and the fraction of pure cooperators decrease as the temptation to defect b increases, as cooperation gets more and more expensive. The fluctuating individuals are present in the network only for a range of intermediate values of b, during which, the cooperation in the system depends almost entirely on them, because there are not pure cooperators anymore.

Regarding the different topologies, we confirm that BA networks can hold higher levels of cooperation than ER networks, even for quite big values of b [8, 9, 18, 28]. As we can see in Fig. 3.2, for random topologies, the average level of cooperation is equal to 1 until it drops quite abruptly around $b = 1.2$, and it disappears almost completely for $b \geqslant 1.8$. For SF networks on the other hand, the cooperation starts decreasing slightly but very soon (for values of $b \gtrsim 1$), but its main drop takes place for higher values (around $b = 1.6$), and, moreover, the cooperation survives for much higher values of the temptation to defect, approximately until $b = 3$. It is interesting to stress again that for values next to $b = 1$, the level of cooperation is $\rho_C = 1$ for ER networks i.e., all the nodes in the system are pure cooperators, but it is slightly lower for SF, since there are already a few fluctuating individuals. Nevertheless, this level of ρ_C will hold on longer before the main fall in SF, while it will drop faster for ER. This fall of ρ_C is present for both topologies, but it is very sharp for ER, so ρ_C drops to zero when $b = 1.3$, while for SF is smoother, allowing the system to keep a small but non-zero value of ρ_C until $b = 2.5$.

3.4 Dipolar Network Model

As we have seen, the asymptotic state of evolutionary dynamics on networks is often not a static equilibrium configuration under the Replicator rule for the update of the strategies. On the contrary, we have shown that there is an asymptotic partition of the graph into three sets, namely, pure cooperators, pure defectors, and fluctuating individuals. This last group experience cycles of invasion by the competing strategies.

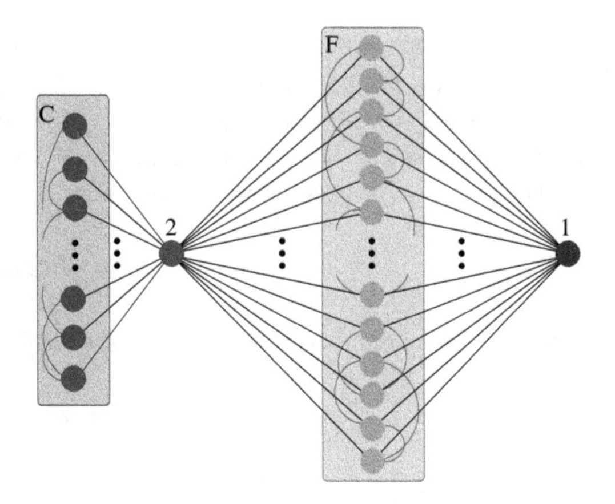

Fig. 3.3 Schematic representation of the Dipolar model network. Nodes 1 and 2 are connected to all nodes in F. Node 2 is also linked to all nodes in C. Connections inside F and C are arbitrary. The colors represent a set of 2^{n_F} different initial configurations. As we usually do, *blue* stands for defector and *red* for cooperator, while *green* means arbitrary strategy

In order to prove the generality of these results, we make a little digression now, and present a model that mimics a local environment of a heterogeneous graph, with simplifications that allow analytical calculations for a better insight. On the other hand, it is perhaps the minimal (though general enough) network model where the partition into PC, PD and F can be rigorously proved, illustrating thus the dynamical organization of cooperation in heterogeneous graphs.

Let's consider the schematic graph in Fig. 3.3, composed of the following elements:

(a) A component F of n_F nodes with arbitrary connections among them.
(b) A node, say Node 1, that is connected to all the nodes in F and has no other links.
(c) A component C of n_C nodes with arbitrary connections among them.
(d) A node, say Node 2, that is connected to all the nodes in F and C, but not to Node 1.

Let's also consider the set of initial conditions defined by: (i) Node 1 is a defector, (ii) Node 2 is a cooperator, and (iii) all nodes in component C are cooperators. Note that this choice allows 2^{n_F} different initial configurations. We now prove that, provided some sufficient conditions (see below), this is an invariant set for the evolutionary dynamics.

If we consider that the nodes are engaged on the Prisoner's Dilemma game, with the specific choices for the parameters of the payoff matrix detailed in Sect. 3.1, then the payoff of a cooperator node i in F is given by:

$$P_i^C = k_i^C + 1 + \epsilon(k_i - k_i^C + 1), \tag{3.7}$$

where k_i is the number of its neighbors in F and $k_i^C \leq k_i$ is the number of those that are cooperators. The payoff of Node 1 is then

$$P_1 \geq (k_i^C + 1)b. \tag{3.8}$$

For the PD game, where $\epsilon < 0$ for the general case, the inequality $P_1 > P_i^C$ always holds, so Node 1 will always be a defector. Thus, a sufficient condition for $P_1 > P_i^C$ is $b > 1 + \epsilon(k_F + 1)$, where k_F ($< n_F$) is the maximal degree in component F, i.e. the maximal number of links that a node in F shares within F.

The payoff of a defector node i in F is

$$P_i^D = (k_i^C + 1)b, \tag{3.9}$$

where k_i^C is the number of its cooperator neighbors in F, while the payoff of Node 2 is

$$P_2 = n_C + n_F\epsilon + n_F^C(1 - \epsilon), \tag{3.10}$$

where $n_F^C \leq n_F$ is the number of cooperators in F. Thus, a sufficient condition for $P_2 > P_i^D$ is $n_C > \text{Int}(b(k_F + 1) - n_F\epsilon)$. With this requisite, Node 2 will always be a cooperator, which in turn implies that all the nodes in the component C will remain always cooperators.

This argument proves that provided the sufficient conditions

$$\begin{aligned} n_C &> \text{Int}(b(k_F + 1) - \epsilon n_F) \\ b &> 1 + \epsilon(k_F + 1) \end{aligned} \tag{3.11}$$

hold, the set of initial conditions defined by (i), (ii), and (iii) is an invariant set: any stochastic trajectory starting in the set remains there. Moreover, as no equilibrium configuration is included in this set, one concludes that no trajectory from this set evolves to an equilibrium configuration. While nodes in C and Node 2 are permanent cooperators, and Node 1 is a permanent defector, nodes in F are forced to fluctuate: at every time step, a defector in F has a positive probability to be invaded by the cooperation strategy, and at the same time, a cooperator in F has a positive probability of being invaded by the defection strategy. In other words, every configuration in the set of initial conditions is reachable (in one time step) from any other, thus it is almost sure that it will be reached (ergodicity).

In any stochastic trajectory starting from the set of initial conditions explained previously, the network is partitioned into three subsets: a set of pure cooperator nodes, a set of pure defector nodes and a set of fluctuating individuals. The fluctuations inside the subpopulation F reflect the competition for invasion among two non-neighboring hubs with fixed opposite strategies in their common neighborhood, a local situation that occurs in heterogeneous networks. It can also be understood as a schematic model for the competition for influence of two powerful superstructural institutions like "mass media", political parties, or lobbies on a target population.

Let's now obtain some exact results for the simplest choice of topology of connections inside the fluctuating set, namely $k_F = 0$. It means that in this case each node in F is only connected to Nodes 1 and 2. Note that the sufficient conditions for fixation of defection at Nodes 1 and 2 are respectively, $b > 1+\epsilon$, and $n_C > b - \epsilon n_F$.

Denoting by $c(t)$ the instantaneous fraction of cooperators in F, the payoffs of Nodes 1 and 2 are

$$P_1 = bcn_F, \quad P_2 = n_C + cn_F + \epsilon(1-c)n_F,$$

and the payoffs of a cooperator node and a defector node in F are respectively

$$P_C = 1 + \epsilon, \quad P_D = b.$$

Then one finds for the one-time-step probability Π_{CD} of invasion of a cooperator node in F, it is to say, the probability of a node in F to change from cooperator to defector

$$\Pi_{CD} = \frac{cb - (1+\epsilon)/n_F}{2\Delta}, \tag{3.12}$$

where $\Delta = \max\{b, b-\epsilon\}$. And on the other hand, using the simplifying notation $A = \epsilon + (n_C - b)/n_F$ and $B = 1 + n_C/n_F$ we get

$$\Pi_{DC} = \frac{A + c(1-\epsilon)}{2\Delta B}, \tag{3.13}$$

for the probability of invasion of a defector node in F, meaning analogously, the probability of changing from defector to cooperator. Note that $A > 0$ because Node 2 can not be invaded.

In this way, the expected fraction of cooperators at time $t+1$ is:

$$c(t+1) = c(t)(1 - \Pi_{CD}) + (1 - c(t))\Pi_{DC},$$

and provided $n_F \gg 1$, the fraction of cooperators c in F evolves according to the differential equation

$$\dot{c} = (1-c)\Pi_{DC} - c\Pi_{CD},$$

which after insertion of Eqs. 3.12 and 3.13 becomes

$$\dot{c} = f(c) \equiv A_0 + A_1 c + A_2 c^2 , \tag{3.14}$$

where the coefficients are

$$\begin{aligned}
A_0 &= \frac{A}{2\Delta B} \\
A_1 &= \frac{1 - \epsilon - A + B(1+\epsilon)/n_F}{2\Delta B} \\
A_2 &= -\frac{1 - \epsilon + bB}{2\Delta B}.
\end{aligned}$$

One can easily check ($A_0 > 0$ and $A_2 < 0$) that there is always one positive root c^* of $f(c)$, which is the asymptotic value for any initial condition $0 \leq c(0) \leq 1$ of Eq. 3.14. Thus, cooperation is never driven to extinction even for large values of the temptation to defect b.

Back to the general case, i.e. arbitrary structure of connections in F, it should be emphasized that the sufficient conditions expressed in Eq. 3.11 do not impose bounds on the network's average connectivity $\langle k \rangle$, that can take arbitrarily large values, independently of the game parameters. This result differs from the bound on $\langle k \rangle$ reported in [6, 10] for different stochastic updating rules in the weak selection limit.

3.5 Distribution of the Strategies Among Connectivity Classes on SF Networks

In order to understand the role of the heterogeneity of SF networks on the asymptotic behavior of the dynamics, and once we have proven the existence of a partition of the nodes into different sets of strategies for a general enough case, we will now proceed to study the fraction of pure cooperators, pure defectors and fluctuating nodes, within every class of connectivity, that we denote by ρ_C^k, ρ_D^k and ρ_F^k, respectively. Note that the total fraction of each type of individuals in the system can be written as:

$$\rho_\alpha = \sum_k P(k)\rho_\alpha^k \tag{3.15}$$

with $\alpha = C, D, F$, and being $P(k)$ the degree distribution. Recall that $\rho_C + \rho_D + \rho_F = 1$, and also $\rho_C^k + \rho_D^k + \rho_F^k = 1$. Thus, in Fig. 3.4 we represent the fraction of pure cooperators and fluctuating nodes as a function of the degree of connectivity of the node and the temptation to defect, b. It can be seen that there are very distinct areas: first of all, for $1 < b \leq 1.7$, the pure cooperators control the system, with values of $\rho_C = 0.9$, while there is only a small fraction of fluctuating strategists, among the nodes with medium or low connectivity. When $1.7 < b < 2$, the pure cooperators decrease to $\rho_C = 0.1$, being set only on the high connectivity nodes, while the fluctuating individuals take over the low classes, up to $k \leq 11$. There is a third region, where the fluctuating nodes invade higher and higher classes of connectivity as b increases, with the pure cooperators still occupying the very high ones (for example, for $b = 2.9$, only the hubs remain being cooperators). Finally, for even higher values of b, ρ_D starts increasing at the expense of ρ_F, but interestingly enough, it does so quite independently of the degree of connectivity. This has to do with the fact that defectors can not take advantage of the heterogeneity of the system, as we will explain in detail next, so this defector invasion for high values of b is consequently independent of the degree of the nodes.

The preferential fixation of pure cooperators in nodes with high degree k when cooperation is very expensive can be understood by the following plausible argument [9, 17, 18]: a necessary though non sufficient condition for a node i to be a pure

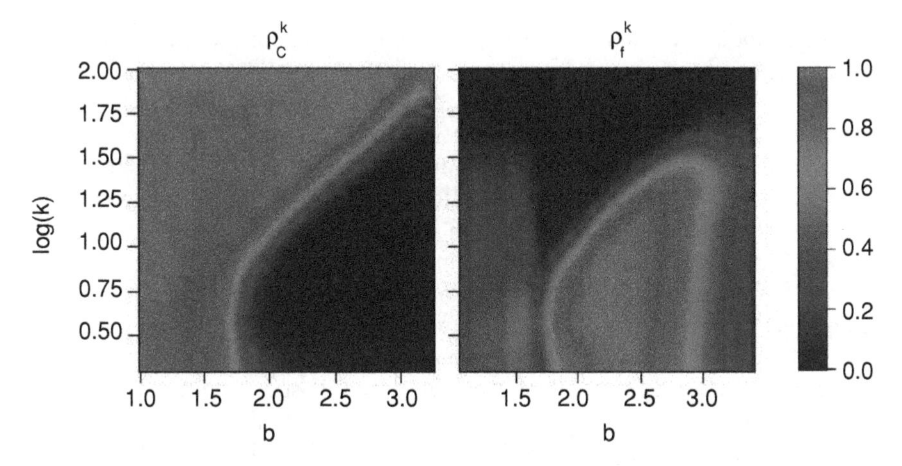

Fig. 3.4 Strategists proportion by classes of connectivity. Color-coded densities of pure cooperator (*Left*) and fluctuating individuals (*Right*) as a function of k and b for BA networks. The size of the networks is $N = 4 \cdot 10^3$ nodes and average connectivity $\langle k \rangle = 4$

cooperator at a given time t is that the number k_i^C of instantaneous cooperators in its neighborhood (i.e., the payoff of i in the current round, since $R = 1$ and $S = 0$) must be greater than the current payoff of any instantaneous defector neighbor j, that is, $k_i^C > bk_j^C$. This condition is clearly favored when the cooperator node i belongs to a high k class and its fluctuating neighbors j belong to lower k classes. This argument is consistent provided that heterogeneous topologies in general either have no degree-degree correlations, so the neighbors of a node of degree k have no preferential degrees, or they are assortative, i.e., neighbors of high degree nodes have preferentially also high degrees. Specifically, SF networks used here, built via preferential attachment using the GM model [25], do have age-correlations, which means that the oldest nodes of a network are usually the hubs, and moreover, they are interconnected, since they formed the initial core of size m_o from which the whole system was grown. This particular feature enhances even more cooperation, so if one destroys such age-correlations, by rewiring the structure and preserving the degree distribution, the average level of cooperation achieved by the system will suffer an important drop, as we will see in some detail in Chap. 5.

The fixation of pure cooperation on hubs is a byproduct of the stabilization of cooperation around them. If we set a cooperator on a hub, it will get very high payoff, because it has very high connectivity, and it will make a lot of its neighbors imitate its strategy. Thus, an all-cooperating area will be created around the hub, from which every cooperator involved will get high benefits too (specially the hub, of course), making this situation very stable. It is to say, the imitation of a successful cooperator hub by its neighbors reinforces its future success, then favoring the fixation of cooperation in highly connected nodes. Nonetheless, if a hub is occupied by a defector, it will get high benefits at the beginning, due to its high connectivity, exploiting all its cooperator neighbors. But this will make more and more of them imitate it, creating an all-defector area around the hub, where nobody will get any benefits at all (recall that a defector against another defector gets $P = 0$). And so the hub will stop getting high payoff too, eventually becoming susceptible of being invaded by a cooperator. In that way, the imitation of a successful defector hub undermines its future success, so that defection cannot take long-term advantage from degree heterogeneity. In a static topology scenario it is impossible for a defector to persist on a hub in the long term. Nonetheless, when dealing with growing heterogeneous structures, a very different picture can arise, as we will see in Chap. 7.

We also want to point out that, as we show on the left panel of Fig. 3.4, for a fixed given value of $b > 2$, ρ_C^k varies rather quickly from 0 to 1 in a small interval of values of k centered around some b-dependent value $k^*(b)$, so that the nodes with degree $k > k^*(b)$ are mostly pure cooperators and those with degree $k < k^*(b)$ are mostly fluctuating (see right panel, $2 < b < 2.9$). In the absence of degree-degree correlations the degree distribution density in the neighborhood of a given node is independent of the node degree, and thus the proportion of cooperators in the neighborhood of a given node is that of the whole network. This implies that the necessary condition for a pure cooperator i, stated previously ($k_i^C > bk_j^C$), becomes $k_i > bk_j$, where j is the fluctuating neighbor of i with highest degree, say $k_j \simeq k^*$. Now, a small increase Δb makes those pure cooperators i fulfilling $(b + \Delta b)k^* > k_i > bk^*$ become fluctuating, so that $\Delta k^* \simeq k^* \Delta b$. From these conditions one concludes that $k^*(b)$ grows exponentially with b, $k^*(b) \propto \exp(b)$. The linear shape of the bright-color line in the $(b, \log k)$ plane at the left panel of Fig. 3.4 for $b > 2$ nicely confirms this prediction, thus supporting the validity of our heuristic argument.

Finally, we want to mention that the invasion process of defectors as the temptation to defect increases on a SF topology could be quite different if we were dealing with structures with a high level of clustering coefficient. As it has been investigated in [29], the existence of a high number of triangular relations within a SF network makes cooperation resilient for even higher values of b on the one hand, but also makes the invasion of defectors quite independent of the degree classes. It is to say, defectors invade homogeneously all the classes of connectivity almost at the same time, which makes the plot $\langle c\rangle(b)$ much sharper.

3.6 Cooperation Times of the Fluctuating Set on SF Networks

We have stated that the fluctuating subpopulation in the dipolar model (see Sect. 3.4) is such that any fluctuating individual has a positive probability of changing strategy in one time step, so that the dynamics is ergodic in the set of all configurations compatible with the partition. This is not necessarily the case in a general heterogeneous network, being perfectly possible that a fluctuating node at a given time has a null one-time-step probability of invasion, but a positive n-time-steps probability for some $n > 1$; thus, ergodicity in the set of configurations compatible with the partition is neither ensured nor discarded.

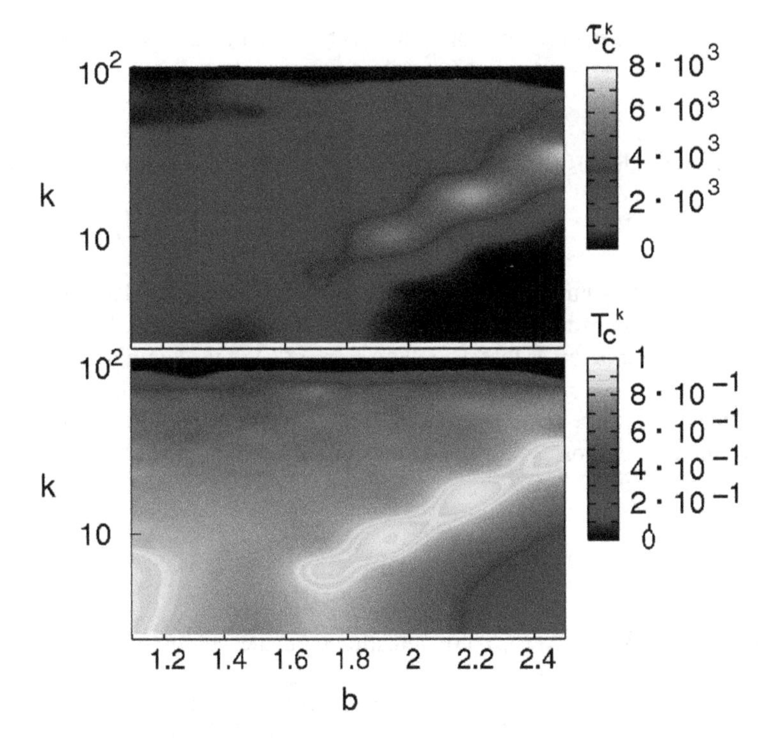

Fig. 3.5 Cooperation times in the fluctuating set. Permanence times τ_C of the cooperation strategy of a fluctuating node (*Top*) and the fraction of time T_C it cooperates (*Bottom*) as a function of the node's degree k and the game parameter b for BA networks and $\epsilon = 0$. The size of the networks is $N = 4 \cdot 10^3$ nodes and average connectivity $\langle k \rangle = 4$

In SF graphs each fluctuating individual is wired to (and then could be invaded by) a different number of fluctuating individuals, and (eventually) pure strategists, so that one should expect that the fraction of time T_C it spends as cooperator differs widely from node to node. The lower panel of Fig. 3.5 shows the average fraction of time T_C^k a fluctuating node of degree k spends cooperating. The average of these quantities $\sum_k P(k)T_C^k$ in the subpopulation F, defines the parameter $\langle T_C \rangle$ that appears in Eq. 3.6, i.e. the average individual contribution of fluctuating nodes to the macroscopic level of cooperation $\langle c \rangle$. To avoid misunderstandings concerning the relative importance of the contribution of connectivity classes to $\langle c \rangle$, it is important to bear in mind both, the power-law dependence of $P(k)$ and the right panel of Fig. 3.4, showing the fraction ρ_F^k of fluctuating nodes inside the class of degree k.

Given that T_C is a proportion of time, it does not provide information on the time scales of the invasion cycles that fluctuating nodes experience. The random variable τ_C (cooperation permanence time) is defined as the time spent as cooperator by a fluctuating node in each cycle. For the dipolar network, when $k_F = 0$, the one-time-step invasion probabilities, Π_{CD} and Π_{DC} (Eqs. 3.12 and 3.13), become time independent in the asymptotic regime. Then one can compute the probability that the cooperation strategy remains for a time $\tau_C \geq 1$ at a fluctuating node, simply as

$$P(\tau_C) = \Pi_{CD}(1 - \Pi_{CD})^{\tau_C - 1}. \quad (3.16)$$

In a similar way, the distribution density $P(\tau_D)$ of defection permanence times is obtained as

$$P(\tau_D) = \Pi_{DC}(1 - \Pi_{DC})^{\tau_D - 1}. \quad (3.17)$$

Thus the density distributions of permanence times for both strategies are exponentially decreasing. For example, at $\epsilon = 0$, i.e. at the border between the PD and the HD game, if one further assumes that the relative size $\mu(F)$ of the component F is large enough, i.e. $\mu(F) \to 1$, and $\mu(C) \to 0$, one obtains that the stationary solution of Eq. 3.14 behaves as $c^* \simeq (b+1)^{-1}$ near the limit $\mu(F) \to 1$. The distribution density $P(\tau_C)$ of the cooperation permanence times of a fluctuating node, as a function of the parameter b is thus

$$P(\tau_C) = (2b + 1)^{-1} \left(\frac{2b + 1}{2b + 2}\right)^{\tau_C}, \quad (3.18)$$

and the distribution density $P(\tau_D)$ of defection permanence times

$$P(\tau_D) = (2b(b + 1) - 1)^{-1} \left(\frac{2b(b + 1) - 1}{2b(b + 1)}\right)^{\tau_D}. \quad (3.19)$$

For SF networks, one expects that the permanence times at the fluctuating nodes show some correlation with the node's degree. The upper panel of Fig. 3.5 represents the average permanence time, τ_C^k, that fluctuating nodes of degree k spend as cooperators as a function of b and k, for observation times of 10^4 generations. We see that cooperation permanence times are strongly correlated with degree: highest τ_C occurs along the line $k^*(b)$ of maximal degree in the fluctuating set.

As we have mentioned before, the heterogeneity of social contacts in SF networks provides local environments where cooperation has a distinctive selective advantage at high degree nodes. This not only enhances the size of the subpopulation where fixation of cooperation occurs, but also enlarges the average total fraction of time of cooperation in the fluctuating subpopulation.

3.7 Microscopic Organization Dynamics of Cooperation

We would like to achieve now a better understanding of the important differences found between the random and the SF topologies, and in order to do that, we will perform a microscopic study of the dynamic organization of the system. First of all, we need to define the concept of cluster or core of nodes for both strategies. A *cooperator cluster* (CC) is a connected component (a subgraph) fully and permanently occupied by cooperator (strategy $s_i = 1$), i.e., by pure cooperators so that $P(s_i(t) \neq 1, \forall t > t_0, \forall i \in CC) = 0$. Analogously, a *defector cluster* (DC) is the subgraph whose elements are pure defectors, namely, when the condition $P(s_i(t) \neq 0, \forall t > t_0, \forall i \in CD) = 0$ is fulfilled. It is easy to see that a CC cannot be in direct contact with a DC, but with a cloud of fluctuating elements that constitutes the frontier between these two cores. Note that a CC is stable if none of its elements has a defector neighbor coupled to more than k^C/b cooperators, where k^C is the number of cooperators linked to the element. Thus, the stability of a CC is clearly enhanced by a high number of connections among pure cooperators, which implies abundance of cycles in the CC. This microscopic structure of clusters is at the root of the differences found in the levels of cooperation for both networks and explains why cooperative behavior is more successful in SF networks than in homogeneous topologies. In fact, as far as loops are concerned, the main difference between the two topologies is that the number of small cycles of length L, N_L, are given by $(\langle k \rangle - 1)^L$ and $(\log N)^L$, respectively [30–32]. Therefore, it is more likely that SF networks develop a CC than ER ones. This has been tested numerically by looking at the probability that at least one cooperator core exists. The results [17] indicate that this probability remains equal to 1 for SF networks even for $b \lesssim 2$ and that it approaches zero for large values of b. On the contrary, for ER networks, the same probability departs from 1 and shows a sudden drop to zero for $b = 2$.

Thus, we will focus now on the number of clusters of cooperators N_{cc} and the number of clusters of defectors N_{cd} for both topologies. In Fig. 3.6 we show the dependence of N_{cc} and N_{cd} with b for ER and BA networks. The first and most relevant result we notice concerns the number of cooperator cores: while for ER graphs N_{cc} there is a wide region of b where there are several clusters of cooperators, for the SF networks the number of cooperator clusters is always 1: no matter the value of b, they always form a single core. We have also verified that the cooperation core in SF networks contains the hubs, which are the ones that hold the whole cluster together, that would otherwise be disconnected. This important difference greatly contributes to the well-known advantage of cooperators in SF networks, compared to ER. Looking at the organization of pure defectors, one can see that there are important differences depending on the topology, too. In ER networks, pure defectors first appear distributed in several clusters that later coalesce to form a single core for values of $b \lesssim 2$, it is to say, before the whole system is invaded by defectors. Conversely for SF topologies, defectors are organized in several clusters, except when they finally occupy the whole system completely. This latter behavior results from the role that hubs play: as they are the most robust against defector's invasion, highly connected individuals survive as pure cooperators until the fraction ρ_C vanishes, thus keeping around them a highly robust cooperator core that loses more and more elements of its outer layer as b increases, until cooperation is finally defeated by defection. In Fig. 3.7 we show the dependence of the number of clusters of defectors N_{cd} as a function of the fraction ρ_D of defectors present in the system (realize that this last magnitude obviously increases with b).

We have summarized in Fig. 3.8 the picture obtained from the analysis performed. Clearly, two different paths characterize the emergence (or breakdown) of coopera-

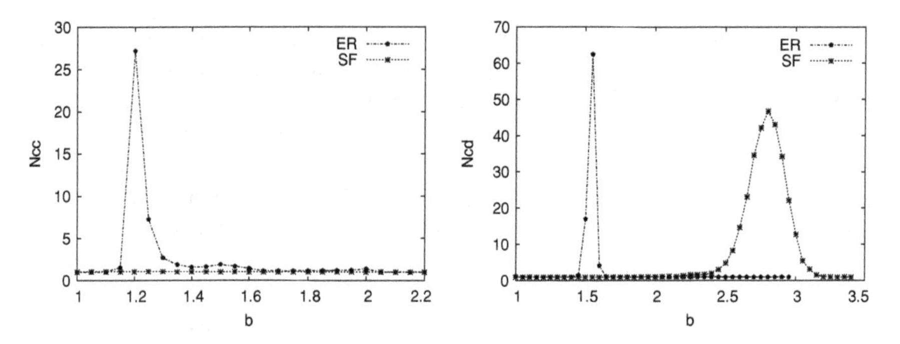

Fig. 3.6 Number of clusters of cooperators (*Left*) and number of clusters of defectors (*Right*) as a function of the parameter b for both ER and BA topologies. The size of the network is $N = 4 \cdot 10^3$ with average connectivity $\langle k \rangle = 4$, and each point shown is the average of 10^3 different realizations of the game and the network

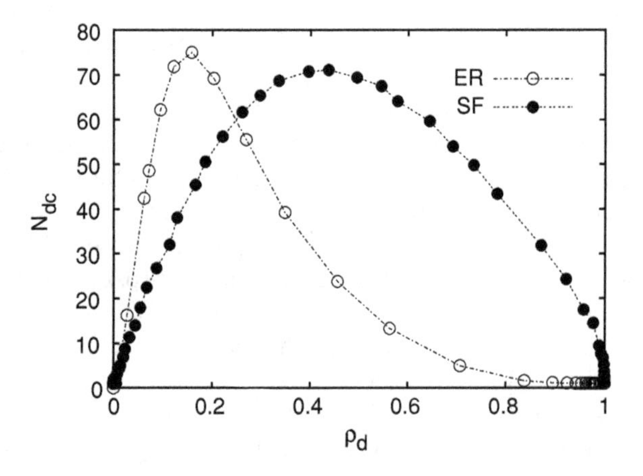

Fig. 3.7 Dependence of the number of clusters of defectors N_{cd} with the fraction of pure defectors in the system ρ_D for both SF and ER topologies (note that, in general, though ρ_D increases with b, the same value of ρ_D for both topologies corresponds to different values of b). The size of the network is $N = 4 \cdot 10^3$ with average connectivity $\langle k \rangle = 4$, and each point shown is the average of 10^3 different realizations of the game and the network

tion. Starting at $b = 1$ all individuals in both topologies are playing as pure cooperators. However, for $b > 1$, the pure cooperative level in SF networks drops below 1 and the population is constituted by pure cooperators forming a single CC, as well as by a cloud of fluctuating individuals. As b is further increased, the size of the cooperation core decreases and some of the fluctuating nodes turn into pure defectors. These defectors are grouped in several clusters around the fluctuating layer (recall that pure cooperators and pure defectors are never in direct contact). For even larger payoffs, the cooperator core is reduced to a small loop tying together a few individuals, among which is highly likely to find the hubs, while the cores of pure defectors gain in size. Finally, pure cooperators and fluctuating elements are invaded

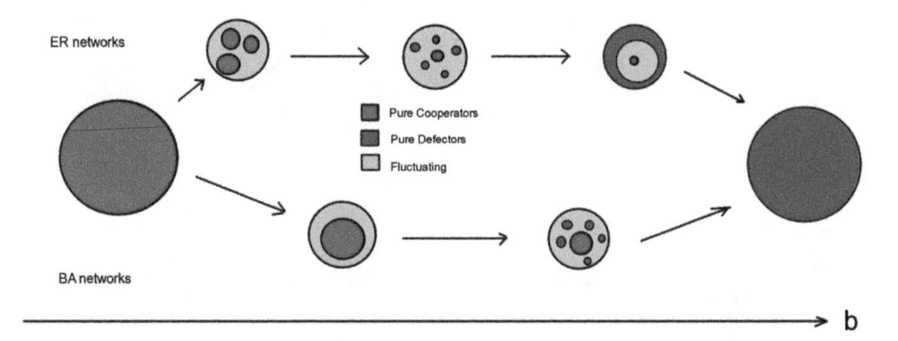

Fig. 3.8 Schematic representation of the different paths from total cooperation to total defection as b increases, for ER and BA topologies

by defectors an a single N-sized defector core is formed. On the contrary, the original N-sized cooperator core survives for higher values of b when it comes to ER graphs. However, when b grows, this cluster splits into several cooperator cores that are in a flood of fluctuating elements. Larger payoffs first give rise to several defector cores that by coalescence form an outer layer that is separated from a single central core of cooperators by individuals of fluctuating strategies. Finally, for $b = 2$, an N-sized defector core takes over.

3.8 Dependence on the Initial Conditions

So far, we have studied the evolution of the PD dynamics on the system starting always from an initial fraction of cooperators equal $\rho_0 = 0.5$, i.e., at the beginning of every simulation, $\rho_0 N$ nodes have been chosen randomly as cooperators on the network, on average. In other words, the initial probability for any node to be a cooperator has been 0.5. Now we want to address the issue of changing this initial cooperation fraction, so it can vary between $0 < \rho_0 < 1$. We want to analyze the possible influence of ρ_0 on the final equilibrium state of the system, for all the range of values of the parameter b and we also want to make a comparison between the two topologies, as usual, ER and SF networks. Besides, it is important to clarify, however, that the distribution of cooperators, given by ρ_0 will still be made randomly among the nodes.

The variation with the game parameter b of the stationary (asymptotic) average level of cooperation, $\langle c\rangle(b)$, for several values of ρ_0, is shown in Fig. 3.9 for ER graphs and BA networks. And as we can see, $\langle c\rangle$ depends on ρ_0 generally speaking, in such a way that increases with it, but this dependence is different for random and SF topologies. When $b \gtrsim 1$, the behavior of $\langle c\rangle$ for both topologies is quite independent from ρ_0, because there is not a big difference between being a cooperator or a defector as far as payoffs is concerned. This is also the case when b is bigger enough to make

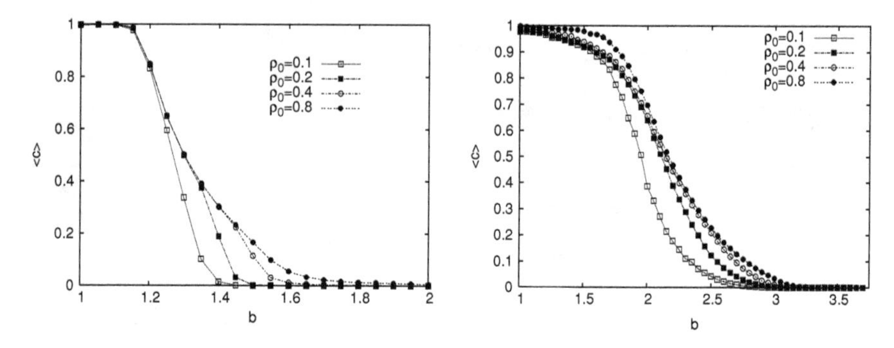

Fig. 3.9 Average cooperation level in ER (*Left*) and SF (*Right*) networks as a function of b for several fixed initial concentrations of cooperators ρ_0 as indicated. The size of the networks is $N = 4 \cdot 10^3$ nodes and average connectivity $\langle k \rangle = 4$. The scale-free network is a BA graph whose $P(k) \sim k^{-3}$. Every point shown is the average of 10^3 different realizations of the game and the network

the whole system defect. But there is a wide range of intermediate values of b where this behavior depends on the heterogeneity of the graph.

In the case of ER networks, different initial concentrations ρ_0 produce a family of curves that mainly differs in their tails, so the larger the value of ρ_0, the slower the decay of $\langle c \rangle$ as b increases (as we will see next, this is in correspondence with the perfect saturation of $\langle c \rangle(\rho_0)$ at fixed b, Fig. 3.10). On the other hand, in BA networks the effects of different initial conditions are appreciated in the whole range of b values. We thus see that heterogeneity not only favors the survival of cooperation, but also makes the value of the average cooperation, at fixed b value, more dependent on initial conditions.

In order to study these differences more thoroughly, we plot these same results as $\langle c \rangle$ versus ρ_0 for several values of the (fixed) parameter b. As it can be seen in Fig. 3.10, $\langle c \rangle$ typically increases with ρ_0 until saturation is reached much before ρ_0 approaches 1. One observes that saturation occurs sooner for smaller values of b. These features are common for both classes of networks. However, some details of the $\langle c \rangle(\rho_0)$ curves are different: first, for ER networks, the departure from zero of $\langle c \rangle(\rho_0)$ occurs, as b increases, only above some b-dependent *threshold* value of the initial fraction of cooperators; on the contrary, for BA networks $\langle c \rangle(\rho_0)$ departs from zero as long as $\rho_0 > 0$, at all values of b inside the coexistence region between both strategies. Second, saturation is more perfect for ER networks, while for BA graphs the plateau in the $\langle c \rangle(\rho_0)$ curve shows a small positive slope. It is interesting to consider these results in the light of those found for the Prisoner's Dilemma in regular square lattices, where the proportion of C and D tends to depend on the starting proportion for relatively small values of b, but for larger b the proportions are essentially independent of the initial configuration [20].

Let's now focus on the relation between the fraction of pure strategists (ρ_C and ρ_D) and the parameter b. As stated in the Sect. 3.2 (and [17]), for any asymptotic trajectory there is a partition of the network into three sets, namely the set PC of pure

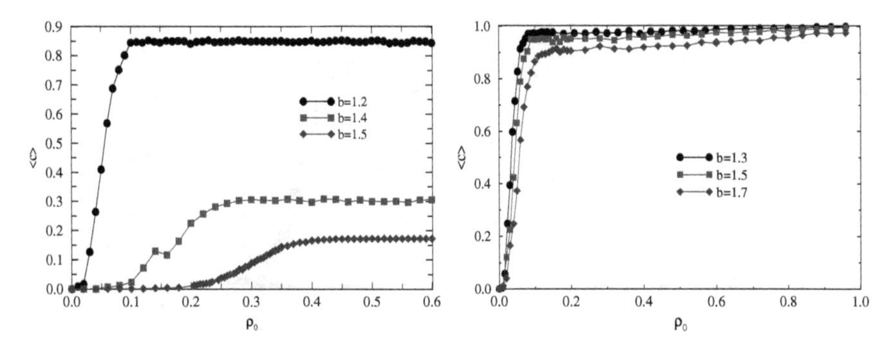

Fig. 3.10 Average cooperation level in ER networks (*Left*) and BA networks (*Right*) as a function of the initial concentration ρ_o for several values of b as indicated. The size of the networks is $N = 4 \cdot 10^3$ nodes and average connectivity $\langle k \rangle = 4$. Each point shown is the average of 10^3 different realizations of the game and the network

cooperator nodes, the set PD of pure defector nodes, and the set F of fluctuating nodes. The behavior of ρ_C and ρ_D versus the game parameter b is plotted in Fig. 3.11 for different initial cooperator concentrations. The first remarkable result is that in ER networks, the density of pure cooperators does not depend on ρ_0 for the *whole* range of b values, in sharp contrast with the above mentioned results for the tails of the average level of cooperation $\langle c \rangle(b)$ (Fig. 3.9). It is worth recalling that, as we have discussed in Sect. 3.6, there are two additive contributions to the average fraction $\langle c \rangle$ of cooperators, namely the measure ρ_C of the set of pure cooperators, and the overall fraction of time $\langle T_C \rangle$ spent by fluctuating nodes as cooperators, weighted by the relative size $\rho_F = \langle \mu(F) \rangle$ of the fluctuating set (see Eq. 3.6). Though the first contribution is, for ER networks, independent of ρ_0, the second one does indeed depend on the initial conditions, as inferred from Fig. 3.9 and the relation $\rho_C + \rho_D + \rho_F = 1$. High initial concentrations of cooperators favor the fluctuating set F at the expense of pure defectors, while the number of nodes where fixation of cooperative strategy occurs remains apparently unaffected. Thus, ρ_C is being mainly determined by the network structural features. For example, in our simulations, for large values of b where ρ_C is very small, we have observed that the pure cooperator nodes form cycles. The fixation of cooperation in these structures is assured if none of their elements is linked to a fluctuating individual that, while playing as a defector, is coupled to more than k_C/b cooperators, where k_C is the number of cooperators attached to the element. The number of such structures is finite in ER graphs, but as soon as their vertexes are occupied by cooperators, they will be immune to defectors invasion.

The right panel of Fig. 3.11 shows the results obtained for BA networks. Regarding the proportion of pure cooperators, one may differentiate two regimes: For $b < 1.7$, there is a moderate dependence of ρ_C on ρ_0, while ρ_C is almost independent of ρ_0 for larger values of b. This behavior correlates well with our observations (Sect. 3.5) on the distribution of strategists inside the degree classes. In the first range of b values, pure cooperators are present in all k-classes and fluctuating individuals are

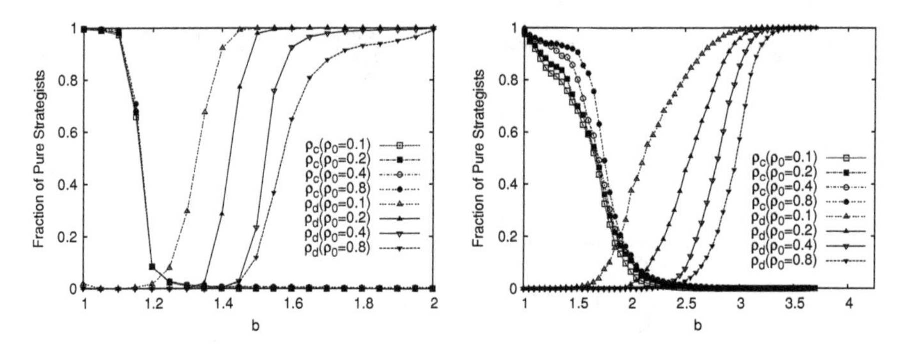

Fig. 3.11 Fraction of pure strategists in ER (*Left*) and SF (*Right*) networks as a function of b and several values of ρ_0. The size of the system is $N = 4 \cdot 10^3$ nodes, with average connectivity $\langle k \rangle = 4$. Every point shown is the average of 10^3 different realizations

almost homogeneously disseminated over low-to-intermediate k classes. However, for $b > 1.7$, there is a b-dependent value of k, say k^*, such that k-classes are fully occupied by pure cooperators if $k > k^*$ while basically no pure cooperators are found in lower k-classes. In the second range of b values, where the degree-strategy correlations are strong, the influence of ρ_0 on the asymptotic proportion of pure cooperators is very small.

As discussed in previous paragraphs, while the proportion of pure cooperators is either independent (ER) or slightly dependent (BA) on initial concentration ρ_0, the measures of the other sets in the partition, F and PD, are indeed more influenced by the initial conditions. The dependence of the fraction of pure defectors ρ_D with ρ_0 for BA and ER networks is qualitatively the same, that is, the proportion of pure defectors is favored (at the expense of the fluctuating set) by a higher initial proportion of defectors. This is consistent with the lack of degree preference (correlation) of pure defectors, which cannot take advantage of degree inhomogeneity: the higher their instantaneous payoff, the more likely they invade neighboring nodes, which has the effect of diminishing their future payoff.

Finally, we analyze the connectedness of the pure strategists sets, as measured by the number of cooperator cores N_{cc}, and defector cores N_{dc}. As we have shown in Sect. 3.7 for BA networks and $\rho_0 = 0.5$, for all values of b where PC is not an empty set, it is connected, i.e. $N_{cc} = 1$. Looking at Fig. 3.12, it can be said that this result turns out to be independent of ρ_0: there is only one cooperator core in BA networks, which contains always the most connected nodes or hubs, for any initial fraction of cooperators. The grouping of pure cooperators into a single connected set PC allows to keep a significant fraction of pure cooperators isolated from contacts with fluctuating nodes. This "Eden of cooperation" inside PC provides a safe source of benefits to the individuals in the frontier, reinforcing the resilience to invasion of the set. Pure defectors, on the contrary, do not benefit from grouping together, and the set PD appears fragmented into several defector cores. Note that for values of $b \simeq 1$, where the set PD is empty, $N_{dc} = 0$, while for very high values of b defection

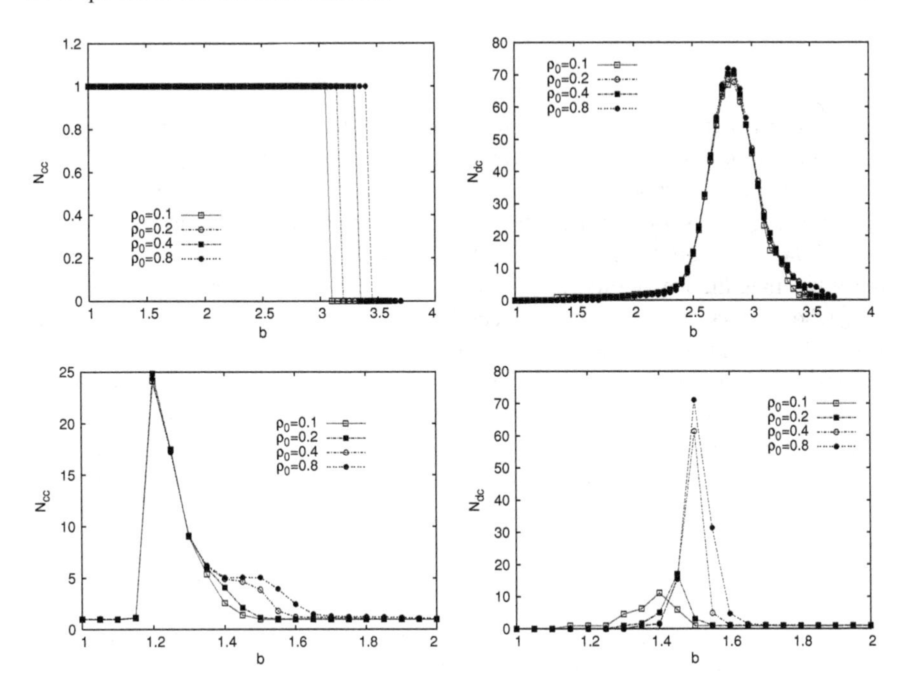

Fig. 3.12 Dependence with b of the number of clusters of cooperators (N_{cc}) and defectors (N_{dc}) for both BA networks (*Top*) and ER graphs (*Bottom*), and for different values of ρ_0. The size of the system is $N = 4 \cdot 10^3$ nodes, with average connectivity $\langle k \rangle = 4$. Every point shown is the average of 10^3 different realizations

reaches fixation in the whole network, so that $N_{dc} = 1$. Thus, $N_{dc}(b)$ must increase first and then decrease to 1. In Fig. 3.12 we show the computed $N_{dc}(b)$ curves for BA networks for several values of ρ_0. It is remarkable that these curves almost collapse, in spite of the fact that the fraction ρ_D of pure defectors does indeed depend on ρ_0 (see Fig. 3.11). This fact suggests that it is the size of the defector clusters, what changes with b, not its number, for the case of BA structures.

In Fig. 3.12 we also show the number of clusters $N_{cc}(b)$ and $N_{dc}(b)$ for ER graphs, and for different fixed values of ρ_0. Regarding the number of cooperator cores, first we notice that the picture described in Sect. 3.7 for the case $\rho_0 = 0.5$ still holds when it comes to other values of the initial proportion of cooperators, it is to say, in general both cooperators and defectors form several unconnected clusters. We also see that except in the small range $1.4 < b < 1.6$, the different curves $N_{cc}(b)$ coincide, in fair agreement with the independence of ρ_C on initial conditions. Note that in the small interval where they do not coincide, the fraction ρ_C of pure cooperators is below 1%, for all values of ρ_0. On the other hand, we see that for higher initial proportion ρ_0 of cooperators, the set PD is more fragmented and also that N_{dc} reaches its maximal values at higher values of b.

3.9 Influence of the Degree of Heterogeneity of the Network

As we established at the beginning of this chapter, we have been comparing the results of the PD dynamics and its microscopical organization for the extreme cases of the GM model, i.e., for random and SF topologies only. Now it is the time to analyze the possible differences for intermediate degrees of heterogeneity. In order to inspect in detail how the results depend on the degree distribution of the network, we monitor the same magnitudes studied previously just for SF and random topologies, but now when the value of α varies between 0 and 1 (we will also include the extreme values, for better understanding). As we have mentioned before, the GM model builds networks with different degree of heterogeneity, depending only on the parameter $\alpha \in [0, 1]$, in such a way that makes the networks less heterogeneous as α increases and approaches 1.

Figure 3.13 shows, from left to the right and from top to bottom, the average level of cooperation $\langle c \rangle$, the density of pure cooperators ρ_C and the density of pure defectors ρ_D as a function of b for several values of α. In this case, the initial distribution of cooperators was set again to $\rho_0 = 0.5$, i.e., at $t = 0$ the nodes have the same probability to cooperate or to defect. The results show that indeed the densities of pure strategists and the average level of cooperation do depend on α. Therefore, the role played by the underlying topology is confirmed: the more homogeneous the graph is, the smaller the level of cooperation in the system for a fixed value of the temptation to defect b. Moreover, the transition for different values of α is absolutely smooth and the systems do not exhibit any abrupt crossover from one kind of behavior ($\alpha = 0$) to the other ($\alpha = 1$).

We have also explored how nodes where strategies have reached fixation are organized into clusters of cooperation and defection as a function of α. Figure 3.14 summarizes our computational simulations for the number of cooperator cores. In this case, we have represented N_{cc} as a function of $(1 - \rho_C)$, that obviously grows with b. We do it this way in order to have a common reference for different values of α until cooperation breaks down, so the comparison is easier. The observed dependence of N_{cc} with α is again smooth and no abrupt change in the behavior of this magnitude occurs. It is worth stressing that as soon as the underlying network departs from the limit $\alpha = 0$ corresponding to a BA scale-free network, the number of CC slightly differs from 1. This means that some realizations give rise to more than one cluster of CC. The probability to have such realizations is very small, but in principle, they are possible. As α is further increased towards one, it is clear that pure cooperators do not organize anymore into a single cluster. We think that this deviation is due to the fact that when $\alpha > 0$ the exponent γ of the underlying network, which still is a scale-free degree distribution, is larger than 3. It is known that this value of γ marks the frontier of two different behaviors when dynamical processes are run on top of complex heterogeneous networks. This is the case, for instance, of epidemic spreading. For $2 < \gamma \leq 3$, the second moment of the degree distribution $P(k)$ diverges in the thermodynamic limit, while it is finite for $\gamma > 3$. As the critical properties of the system are determined by the ratio between the first

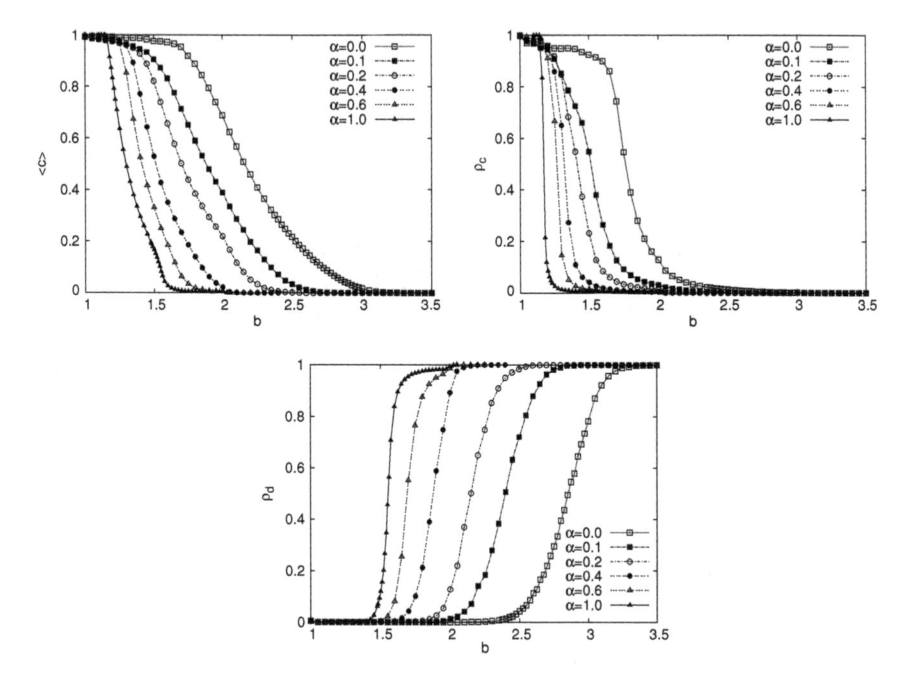

Fig. 3.13 Average level of cooperation (*Top left*) and densities of pure cooperators (*Top right*) and pure defectors (*Bottom*) as a function of b for different values of α ($\alpha = 0$ corresponds to a BA network while $\alpha = 1$ generates an ER graph). In this case, the networks are made up of $N = 2 \cdot 10^3$ nodes and average connectivity $\langle k \rangle = 4$. Every point shown is the average of 10^3 different realizations

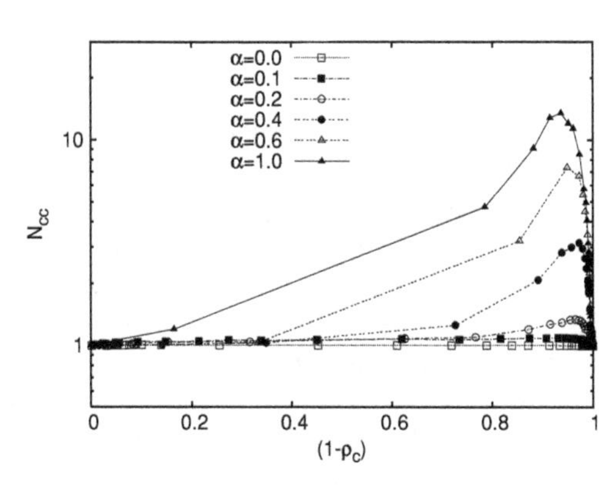

Fig. 3.14 Number of cooperator cores for different networks defined by the value of α as a function of the density of nodes that are not pure cooperators $1-\rho_C$. The networks are made up of $N = 2 \cdot 10^3$ nodes and average connectivity $\langle k \rangle = 4$, and each point shown is the average of 10^3 different realizations of the game and the network

(that remains finite for $\gamma > 2$) and the second moment, the divergence of the latter when $N \to \infty$ and $2 < \gamma \leq 3$, makes the epidemic threshold null. On the contrary, when the process takes place in networks whose $\gamma > 3$, the epidemic threshold is recovered, although no singular behavior is associated to the critical point [33, 34]. We expect that a similar phenomenology is behind the results shown in Fig. 3.14.

3.10 Conclusions

In this chapter we have studied the influence of the topology on the dynamics, specifically, the differences between ER and SF networks when implementing the Prisoner's Dilemma on top of them. On the one hand, we have measured the mean levels of cooperation as a function of the one free parameter of the game, the temptation to defect, b, as well as the dependence with the initial proportion of cooperators present on the system, and we have also checked the distribution of the cooperation among the connectivity classes, for the SF networks. On the other hand, we have shown and analytically proved that there is always a partition of the network into three different sets of individuals, as far as strategies are concerned, and we have also determined that two different patterns of cooperative behavior can be clearly identified, depending on the underlying structure.

We have found that the evolution of cooperation in complex topologies shows a very rich structural and dynamical behavior. For values of the temptation to defect b close to one, ER networks outperform SF topologies, but the presence of hubs and the relative abundance of small loops in SF networks revert the behavior of the level of cooperation for intermediate to large values of the payoffs. The reason why SF networks can foster much higher levels cooperation than ER, even when the temptation to defect makes it very expensive, is that heterogeneous populations offer to the cooperative strategy the opportunity of evolutionary mechanisms of positive feedback, making cooperation the fittest overall strategy, in spite of not being the best reply to itself in the one-time step scenario. Besides, we have found that regardless of the topology and even the values of the parameters of the model, there are always three different classes of individuals according to their asymptotic strategies: the set of pure cooperators PC, pure defectors PD and fluctuating individuals F, and we have developed a simple but very useful analytical model that mimics the competition for invasion of two highly connected nodes in order to prove the existence of this partition of the network in a general case.

Regarding the microscopic organization of the system, we have found important differences between ER and SF: we have measured the number of clusters of cooperators, and shown that, while in SF networks cooperators form always one single cluster (its relative size depending obviously on the value of the temptation to defect), in homogeneous topologies they form several disconnected clusters, and therefore they are 'an easy target' for the attacks of the defectors. Nonetheless, the number of clusters of defectors is always more than one, in general, for both ER and SF networks.

Here, we have also shown that the enhancement of cooperation due to the heterogeneity of the pattern of connections among agents is robust against variations of the initial conditions (meaning different initial concentrations of cooperators, ρ_0, always randomly distributed on the population). While both the measure of the cooperator set PC where cooperation reaches fixation, and its connectedness properties are either independent or only slightly dependent on ρ_0, the measure of the fluctuating set F and the defector set PD where defection is fixed show a clear dependence on the initial conditions, for defection cannot profit from degree heterogeneity. On the other hand, the characteristics of the asymptotic evolutionary states of the PD analyzed here, show a smooth variation when the heterogeneity of the network of interconnections is one-parametric tuned from Poissonian to scale-free, demonstrating a strong correlation between heterogeneity and cooperation enhancement.

Finally, though the numerical results presented here correspond mostly to network sizes $N = 4 \cdot 10^3$, we have studied also larger networks, up to $N = 10^4$, with no qualitative differences in the results. The increase of network size, while keeping constant the average degree $\langle k \rangle$, turns out to be beneficial for cooperation, due to the fact that it has the effect of increasing the maximal degree, and thus the range of degree values. This further confirms how efficiently cooperation takes advantage of degree heterogeneity.

References

1. W. Hamilton, J. Theor. Biol. **7**, 1 (1964).
2. R. Axelrod and W. Hamilton, Science **211**, 1390 (1981).
3. M. Nowak and K. Sigmund, Nature **437**, 1291 (2005).
4. J. Hofbauer and K. Sigmund, *Evolutionary games and population dynamics*. (Cambridge University Press, Cambridge, UK, 1998).
5. J. Hofbauer and K. Sigmund, Bull. Am. Math. Soc. **40**, 479 (2003).
6. M. Nowak, Science **314**, 1560 (2006).
7. M. A. Nowak and R. M. May, Nature **359**, 826 (1992).
8. F. C. Santos and J. M. Pacheco, Phys. Rev. Lett. **95**, 098104 (2005).
9. F. C. Santos, F. J. Rodrigues, and J. M. Pacheco, Proc. Biol. Sci. **273**, 51 (2006).
10. H. Ohtsuki, E. L. C. Hauert, and M. A. Nowak, Nature **441**, 502 (2006).
11. G. Abramson and M. Kuperman, Phys. Rev. E **63**, 030901(R) (2001).
12. V. M. Eguíluz, M. G. Zimmermann, C. J. Cela-Conde, and M. San Miguel, American Journal of Sociology **110**, 977 (2005).
13. T. Killingback and M. Doebeli, Proc. R. Soc. Lond. **263**, 1135 (1996).
14. G. Szabó and G. Fáth, Phys. Rep. **446**, 97 (2007).
15. A. Szolnoki, M. Perc, and Z. Danku, Physica A **387**, 2075 (2008).
16. J. Vukov and G. S. A. Szolnoki, Phys. Rev. E **77**, 026109 (2008).
17. J. Gómez-Gardeñes, M. Campillo, L. M. Floría, and Y. Moreno, Phys. Rev. Lett. **98**, 108103 (2007).
18. F. C. Santos and J. M. Pacheco, J. Evol. Biol. **19**, 726 (2006).
19. K. Lindgren and M. Nordahl, Physica D **75**, 292 (1994).
20. M. Nowak, S. Bonhoeffer, and R. May, Proc. Natl. Acad. Sci. USA **91**, 4877 (1994).
21. J. Maynard Smith, *Evolution and the Theory of Games*. (Cambridge University Press, Cambridge, UK, 1982).
22. H. Gintis, *Game theory evolving*. (Princeton University Press, Princeton, NJ, 2000).
23. C. Hauert and M. Doebeli, Nature **428**, 643 (2004).
24. C. P. Roca, J. A. Cuesta, and A. Sánchez, Phys. Rev. E **80**, 046106 (2009).
25. J. Gómez-Gardeñes and Y. Moreno, Phys. Rev. E **73**, 056124 (2006).
26. P. Erdős and A. Renyi, Publicationes Mathematicae Debrecen **6**, 290 (1959).
27. A. Barabási and R. Albert, Science **286**, 509 (1999).
28. F. C. Santos, J. M. Pacheco, and T. Lenaerts, Proc. Natl. Acad. Sci. USA **103**, 3490 (2006).
29. S. Assenza, J. Gómez-Gardeñes, and V. Latora, Phys. Rev. E **78**, 017101 (2008).
30. G. Bianconi and A. Capocci, Phys. Rev. Lett. **90**, 078701 (2003).

31. G. Bianconi and M. Marsili, J. Stat. Mech. p. P06005 (2005).
32. E. Marinari and R. Monasson, J. Stat. Mech. p. P09004 (2004).
33. R. Pastor-Satorras and A. Vespignani, Phys. Rev. E **63**, 066117 (2001).
34. Y. Moreno, R. Pastor-Satorras, and A. Vespignani, European Physical Journal B **26**, 521 (2002).

Chapter 4
Other Games on Static Complex Networks

In the last chapter, we have been discussing in some detail the dynamics and microscopical organization of the the so-called *weak* Prisoner's Dilemma Game [1] on static complex networks, where the payoff for a cooperator against a defector was fixed to $S = 0$ (strictly speaking, for this value of S, we are really at the border between the Prisoner's Dilemma game and the Hawks and Doves -HD- game). In this chapter we want to address very briefly the issue of other evolutionary games on graphs.

Given the usual payoff matrix for the 2×2 games:

$$\begin{array}{cc} & \begin{array}{cc} C & D \end{array} \\ \begin{array}{c} C \\ D \end{array} & \begin{pmatrix} R & S \\ T & P \end{pmatrix} \end{array} \tag{4.1}$$

and once we have fixed $R = 1$ and $P = 0$, we have four different games, depending on the relative ordering of the parameters (the first three of which are interesting well-known social dilemmas) [2–4]:

- The Stag Hunt game [5, 6], with $R > T > P > S$, is a coordination game and both strategies are strict Nash equilibria. Players prefer mutual defection to unilateral cooperation ($S < P$), resulting in an intrinsic fear of individuals to cooperate.
- In the Hawks and Doves (or Snow Drift or Chicken) game [7–14], with $T > R > S > P$, the players are referred to as greedy, since they prefer unilateral defection to mutual cooperation ($T > R$). This is an anti-coordination game, because the best strategy for an individual is the opposite to its opponent's.
- In the Prisoner's Dilemma game, for which $T > R > P > S$, and where both tensions described above are incorporated at once, is the most difficult situation for cooperation to arise.
- In the Harmony game, for which $R > S > T > P$, mutual cooperation is the best option. Thus, this game does not represent a very interesting case of study for us.

J. Poncela Casasnovas, *Evolutionary Games in Complex Topologies*, Springer Theses,
DOI: 10.1007/978-3-642-30117-9_4, © Springer-Verlag Berlin Heidelberg 2012

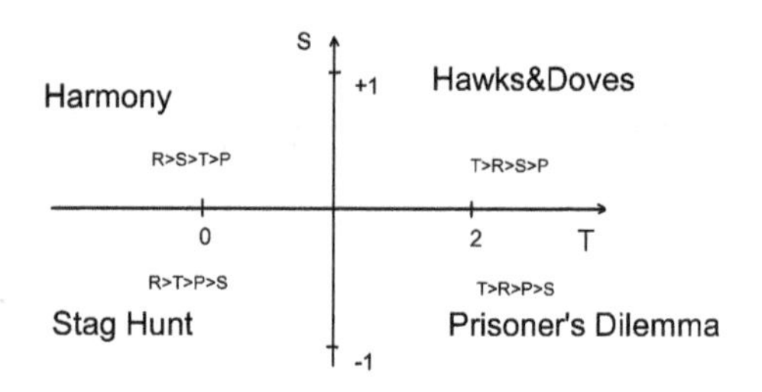

Fig. 4.1 Schematic representation of the different games on the T–S plane

On Fig. 4.1 we sketch the disposition of all of them on the T–S plane.

As we explained in Sect. 2.2, Evolutionary Game Theory predict that cooperation can not survive when playing the Prisoner's Dilemma game on well-mixed populations, whereas there is an interior equilibrium in the Hawks and Doves game, so the system will end up in a situation where a certain proportion of both strategies is present. Nonetheless, we also know of the important differences introduced by the topology on the outcome of the weak PD game. Now we will study and compare the cases of a general Prisoner's Dilemma game and the Hawks and Doves game on complex networks. Our approach will be very similar to the one used in Chap. 3, it is to say, we will study the asymptotic equilibrium state of the system, given by the average level of cooperation, when engaged in general PD game and HD respectively, on top of a complex network, and comparing homogeneous and heterogeneous topologies. Then we will focus on the partition of the graph into several sets, and also on the number and distribution of clusters of the different strategies. Finally, we will look into the level of cooperation among the different connectivity classes for SF topologies. On the one hand, we want to study a more generalized Prisoner's Dilemma game, it is to say, situations with other values of the payoff parameter $S < 0$, to test the results found in the preceding chapter. And on the other hand, we will analyze the behavior of the system when playing another 2×2 game, namely the Hawks and Doves game, and we will compare the outcome of the game for both scenarios.

We have used here the same dynamic rule as in the previous case of the weak Prisoner's Dilemma (Chap. 3), it is, at each time step, every node i of the system plays with its nearest k_i neighbors, as dictated by the underlying network, and accumulates the payoffs P_i obtained during that round. Then, individuals are allowed to synchronously change their strategies by comparing the payoffs they accumulated in the previous generation with the one obtained by a randomly chosen neighbor j. In this way, if $P_i > P_j$, player i keeps the same strategy for the next round of the game, when it will play again with all its neighborhood. On the contrary, whenever $P_j > P_i$, i adopts the strategy of j with probability [11, 15–19]:

$$\Pi_{i \to j} = \beta(P_j - P_i) \tag{4.2}$$

where β^{-1} =max$\{k_i, k_j\}\Delta$, and Δ is the maximum possible difference between the parameters of the payoff matrix, it is to say $\Delta = T$ (given that $S \geq 0$ for the

Hawks and Doves game). For the *weak* Prisoner's Dilemma studied before, it was also $\Delta = T$, but for a more general case of the game, with $S < 0$, it will be $\Delta = T - S$. This probability is proportional to the difference of payoffs of the nodes involved, and it is normalized by the maximum payoff a node can get. It is important to keep in mind that, though it is stochastic, the Replicator-like rule does not allow the adoption of irrational strategy, *i.e.* $\Pi_{i \to j} = 0$ whenever $P_j \leq P_i$. In other words, a node will never adopt the strategy of a neighbor whose payoff was worse than its own in the previous round of the game.

As we did in Chap. 3, the dynamics evolves on top of ER [20] or BA [21] networks, *i.e.* strategists are located on the vertices of a fixed graph that dictates the pattern of social interactions of the population. The size of the system is $N = 4 \times 10^3$ nodes, and the average connectivity of the networks is $\langle k \rangle = 4$. The initial strategy of each one of the N nodes in the system is randomly set, with a probability of being a cooperator equal to $\rho_0 = 0.5$ and then the dynamics starts. We let the system evolve for 5×10^3 time steps or generations, after which we check whether the equilibrium has been reached. As usual, we observe the system during a time window of 10^3 generations. If the slope of $C(t)$ is smaller than 10^{-2}, then we consider that the equilibrium has been reached. Otherwise, we let the system evolve for another 5×10^3 more generations, after which we evaluate it again. The results that we show are usually the average of 10^3 different realizations of networks and initial conditions, except we state otherwise.

4.1 Average Level of Cooperation, and Fractions of Pure Strategists and Fluctuating Individuals

As we have already mentioned, we set $R = 1$ and $P = 0$, and explore the dynamics for a continuum of values of S and T. So the figures we present next will include a comparison of the general Prisoner's Dilemma game ($S < 0$) and the Hawks and Doves game ($S > 0$) at once.

The first result we present is that the asymptotic existence of the partition of the networks into pure strategists and fluctuating individuals (see Sect 3.3) is a general result for the games studied in this chapter. In Fig. 4.2 we show the color-coded average level of cooperation reached for the system after the transient period, as well as the fractions of pure strategists and fluctuating individuals for ER topologies. On the other hand, in Fig. 4.3 we represent the same magnitudes for BA networks.

We also confirm that the dependence of the dynamics on the parameter S is smooth: there is not an abrupt change of behavior around the line $S = 0$. Or in other words, we were entitled to use the weak Prisoner's Dilemma, fixing and eliminating the parameter S, instead of using a more strict version of the game, because the results do not change drastically with small variations around $S \lesssim 0$.

For a fixed value of the temptation to defect T and for both topologies, the more expensive being a cooperator against a defector gets (*i.e.*, S going from positive to

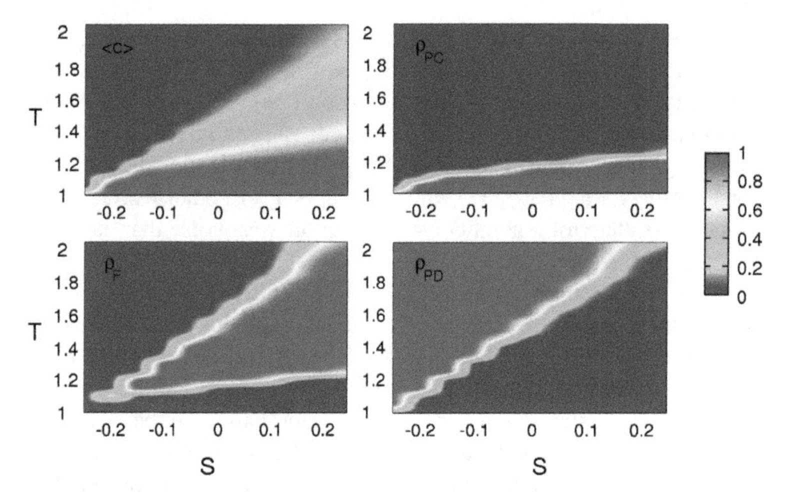

Fig. 4.2 Color-coded average level of cooperation (*Top left*), fraction of pure cooperators (*Top right*), fluctuating individuals (*Bottom left*) and pure defectors (*Bottom right*) for ER networks

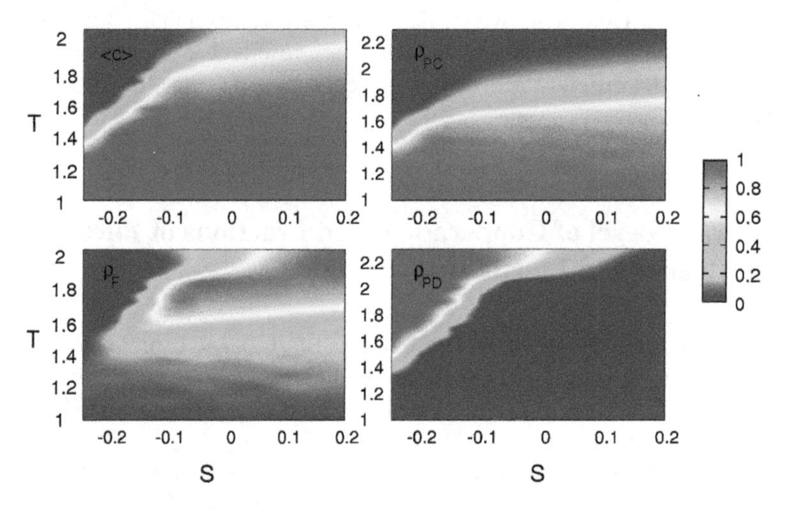

Fig. 4.3 Color-coded average level of cooperation (*Top left*), fraction of pure cooperators (*Top right*), fluctuating individuals (*Bottom left*) and pure defectors (*Bottom right*) for BA networks

negative values), the lower the average level of cooperation. On the other hand, the fraction of pure cooperators ρ_C does not show a strong dependence with S neither for BA nor ER networks. Pure defectors take over the whole system for a wide range of the parameters studied (half of the $S - T$ plane represented, on the ER case). As expected, the value of T for which all the nodes are pure defectors decreases as S does so, since lower values of S or higher values of T make cooperation more expensive. The most important result is that there are not always fluctuating players present on the system for any given value of the parameters. Obviously, if $S < 0$ and T is high enough, all nodes are pure defectors, and if T is low, all individuals act as pure cooperators, no matter what the value of S is. However, there is also an intermediate area of parameters for which the fluctuating nodes occupy almost the entire system, being responsible for the maintenance of the average level of cooperation shown on the system.

If we look at panels in Fig. 4.2, we observe that the frontier between PC and F is almost S-independent, but the frontier between F and PD does depend on the parameter S. This makes that the transition in T from total cooperation to total defection also S-dependent: for high values of S, this transition is smooth, while for negative values of S it is quite sharp, suggesting an almost immediate conversion of the nodes of the system from PC to PD.

Regarding the influence of the topology, as one could expect, both the average level of cooperation and the fraction of pure cooperators are higher for BA than for ER networks. We also see that the fraction of fluctuating individuals (when present) is larger in ER networks, and the limits of the area for which they are present are more clearly drawn in this case.

4.2 Number of Clusters of Cooperators and Defectors

Using the same definition of Cluster presented in the Sect. 3.7, we consider a cooperator cluster (CC) as a connected component (subgraph) fully and permanently occupied by the cooperator strategy $s_i = 1$, *i.e.* composed of pure cooperators so that $P(s_i(t) \neq 1, \forall t > t_0, \forall i \in CC) = 0$. Analogously, a defector cluster (DC) is the subgraph whose elements are pure defectors, that is, a subgraph where the condition $P(s_i(t) \neq 0, \forall t > t_0, \forall i \in DC) = 0$ is fulfilled.

In Fig. 4.4 we show the number of clusters of cooperators N_{cc} and defectors N_{dc} as a function of T for several discrete values of the parameter S and for both ER and BA topologies in both the Hawks and Doves ($S > 0$) and the general Prisoner's Dilemma game ($S < 0$). As we can see, once again the general result obtained previously for the weak Prisoner's Dilemma (Sect. 3.7) holds in these scenarios: while cooperators form several clusters on ER topologies, for the BA networks, as long as cooperators survive in the system, they remain together forming one single cluster which always includes most of the higher connected nodes, making thus the system much stronger against the attacks of the defectors. Those, on the other hand, always form more than one cluster, in general, on both random and scale-free networks. This aspect of the microscopic organization of the strategies in the system for the Hawks and Doves and the general Prisoner's Dilemma is not very surprising, since it is proven to be basically due to the underlying topology, but it definitely affirms the robustness of the results presented in Sect. 3.7, and highlights the differences between homogeneous and heterogeneous networks.

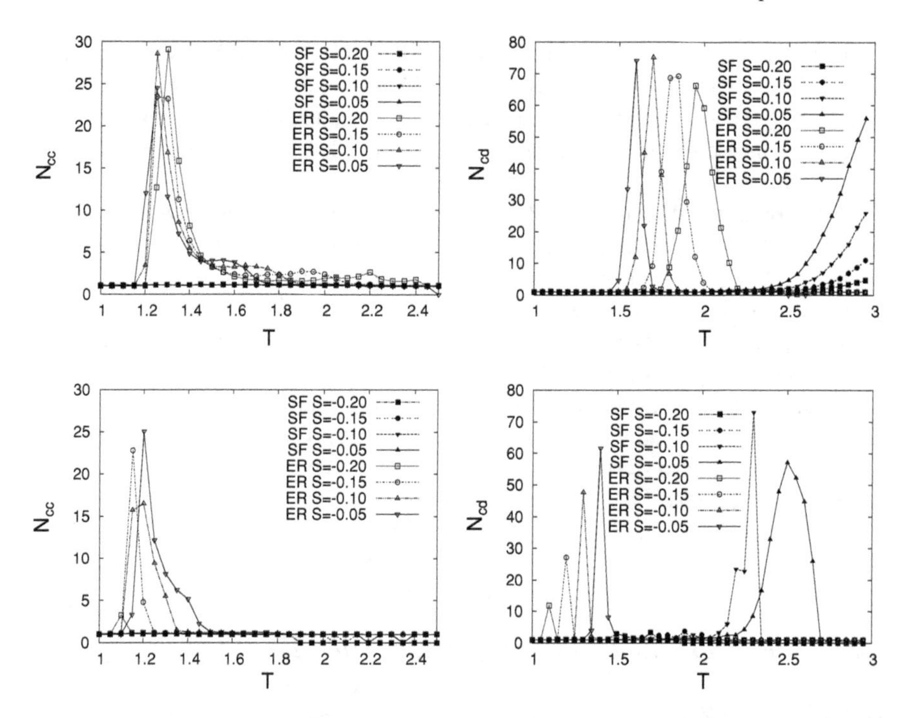

Fig. 4.4 Number of clusters of cooperators N_{cc} and defectors N_{cd} as a function of T for the Hawks and Doves game (*Top panels*) and the general Prisoner's Dilemma case (*Bottom panels*), for both ER (*empty symbols*) and BA (*full symbols*) topologies. The size of the networks is $N = 4000$ nodes and average connectivity $\langle k \rangle = 4$. Every point shown is the average of 5×10^2 values

4.3 Distribution of the Cooperation Among the Degrees of Connectivity

We can study the role of heterogeneity on the dynamics of both PD and HD games by plotting the probability of a node with degree k of being a cooperator, ρ_C^k, in a similar way as we did in Sect. 3.5. Recall that the total fraction of pure cooperators in the system can be written as:

$$\rho_C = \sum_k P(k)\rho_C^k \tag{4.3}$$

with $P(k)$ being the degree distribution, and where the relations $\rho_C + \rho_D + \rho_F = 1$ and $\rho_C^k + \rho_D^k + \rho_F^k = 1$ are fulfilled. As one can see in Fig. 4.5, when T is small enough, all nodes are cooperators, regardless of their connectivity, but as T increases, nodes with intermediate degree are less likely to be cooperators, while the higher classes remain as cooperators until the value of T is such that level of cooperation vanishes in the system completely. This is in perfect agreement with the results

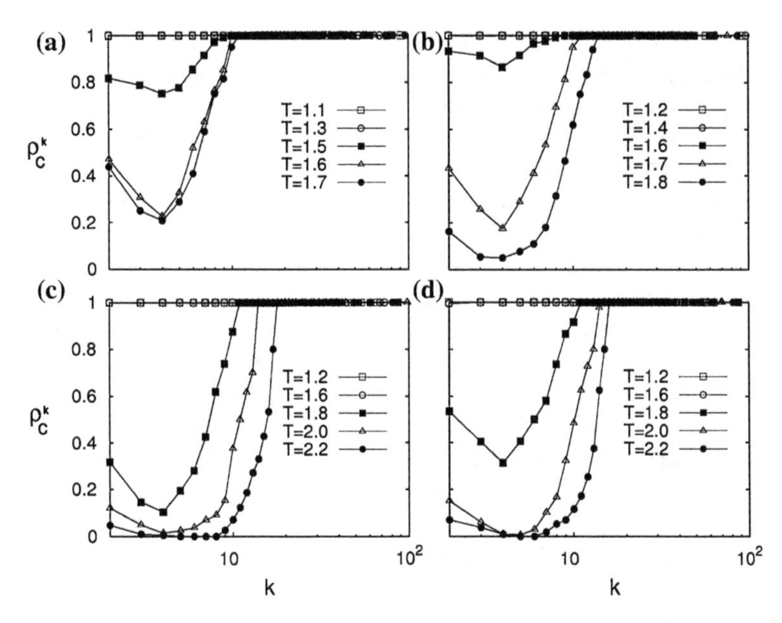

Fig. 4.5 Probability of finding a Pure Cooperator of degree k in SF networks for different values of the parameter T. **a** S $= -0.2$ and **b** S $= -0.1$ correspond to Prisoner's Dilemma scenarios, while **c** S $= 0.1$ and **d** S $= 0.2$ are Hawks and Dove situations. The networks have $N = 2 \times 10^3$ nodes

found for the *weak* Prisoner's Dilemma [17, 22, 23], and shown in Sect. 3.5. As we commented in detail in that section, the reason why the cooperation can survive for SF topologies is because of the existence of the hubs, which are interconnected, play as cooperators, and surround themselves by more cooperators, creating a nice environment of cooperation (or 'Eden') where other cooperator nodes with lower degree can get benefits from it, and resist the attacks of defectors. On the other hand, defectors can not take advantage of the heterogeneity of the network, because they are not stable in the long term when set in a hub.

4.4 Conclusions

In this chapter we wanted to check whether or not some of the important previous results exposed in Chap. 3 for the weak Prisoner's Dilemma (it is, when $S = 0$) still hold for the general case of the game, and even for other two-strategy game, specifically the Hawks and Doves.

As we have seen, given the payoff matrix of the game, the parameter ordering for the Prisoner's Dilemma is $T > R > P > S$, while for the Hawks and Doves game, it is $T > R > S > P$. So, although in both cases players prefer unilateral defection to mutual cooperation, the difference between them is that in the first case, the worst strategy is to cooperate against a defector, while in the second setting, it is to mutually defect. As usual, we have fixed the parameters $P = 0$ and $R = 1$, so to have two free parameters, the temptation to defect, T and the sucker's payoff, S. In this way, for a fixed value of $T > 1$, if $S < 0$, we are playing Prisoner's Dilemma, while if $S > 0$, we are playing Hawks and Doves. Cooperation gets more expensive every time T increases or if S decreases. On the general Prisoner's Dilemma (meaning, for values of $S < 0$ strictly, instead of the weak limit, $S = 0$), we have checked that the dependence with the parameter S is smooth, there are no abrupt changes, but nonetheless, there are some differences. In particular, for a fixed value of the temptation to defect, the more negative S gets, the more expensive the cooperation is, so both the mean value of cooperation, $\langle c \rangle$, and the level of pure cooperators, ρ_C, decrease. And also the level of fluctuating individuals, ρ_F, drops remarkably, while obviously, the level of pure defectors, ρ_D, increases. In this situation, since the levels of F are low, the transition from pure cooperation to pure defection as T increases is quite sharp. On the other hand for Hawks and Doves ($S > 0$) the cooperation is less expensive than for the Prisoner's Dilemma for the same value of T, so both the mean value of cooperation $\langle c \rangle$, and the fraction of pure cooperators ρ_C are obviously higher than in the Prisoner's Dilemma scenario, and the level of fluctuating individuals, ρ_F, is also much higher. In fact, when $S > 0$, there is a wide region of the $S - T$ plane where fluctuating individuals clearly take over the entire system, and this makes the transition from pure cooperation to pure defection as T increases smooth.

Regarding the influence of the underlying topology, we can confirm that the heterogeneity of the network always favors the cooperation for both games. Thus, $\langle c \rangle$ and ρ_C are much higher for SF than for ER networks, while the fluctuating and pure defectors are less present on heterogeneous systems. We have checked the microscopic organization of the cooperation on the system as well, and we have found that the results shown in Sect. 3.7 still hold both for the general Prisoner's Dilemma case and the Hawks and Doves: while for SF topologies, cooperators organize into just one single cluster, for ER they form several. Thus, in the first case the system can hold much higher levels of cooperation even when it is very expensive (for high values of T or negative values of S). On the other hand, the defectors always organize into several clusters, in general, regardless the underlying topology.

Finally, if we look at the distribution of the cooperation across the connectivity classes in SF networks, we can see that, as we have proved previously for the weak Prisoner's Dilemma case, when cooperation is not expensive ($T < 1.5$), practically the whole system plays as a cooperator, but when it gets more expensive, the defectors start taking over the medium classes, while the high classes remain unconquered as long as cooperation can survive. This hierarchical organization is preserved for all the values of S explored.

To summarize, in this chapter, we have proven the robustness and strength of the important results previously shown in Chap. 3. We have proved that all of them hold for a wide range of parameters, specially the important differences regarding the topology and the microscopic organization of the system.

References

1. G. Szabó and G. Fáth, Phys. Rep. **446**, 97 (2007).
2. M. Nowak and K. Sigmund, Nature **437**, 1291 (2005).
3. F. C. Santos, J. M. Pacheco, and T. Lenaerts, PLos Comput. Biol. **2**(**10**), e140 (2006).
4. J. Maynard Smith, *Evolution and the Theory of Games.* (Cambridge University Press, Cambridge, UK, 1982).
5. L. E. Blume, Games and Economic Behavior **5**, 387 (1993).
6. G. Ellison, Econometrica **61**, 1047 (1993).
7. F. C. Santos, J. M. Pacheco, and T. Lenaerts, Proc. Natl. Acad. Sci. USA **103**, 3490 (2006).
8. C. P. Roca, J. A. Cuesta, and A. Sánchez, Phys. Rev. E **80**, 046106 (2009).
9. T. Killingback and M. Doebeli, Proc. R. Soc. Lond. **263**, 1135 (1996).
10. M. Tomassini, L. Luthi, and M. Giacobini, Phys. Rev. E **73**, 106 (2006).
11. C. Hauert and M. Doebeli, Nature **428**, 643 (2004).
12. M. Sysi-Aho, J. Saramaki, J. Kertész, and K. Kaski, European Physical Journal B **44**, 129 (2005).
13. L. Zhong, D. Zheng, B. Zheng, C. Xu, and P. Hui, Europhys. Lett. **76**, 724 (2006).
14. A. Kun, G. Boza, and I. Scheuring, Behav. Ecol. **17**, 633 (2006).
15. H. Gintis, *Game theory evolving.* (Princeton University Press, Princeton, NJ, 2000).
16. F. C. Santos and J. M. Pacheco, Phys. Rev. Lett. **95**, 098104 (2005).
17. F. C. Santos, F. J. Rodrigues, and J. M. Pacheco, Proc. Biol. Sci. **273**, 51 (2006).
18. J. Hofbauer and K. Sigmund, *Evolutionary games and population dynamics.* (Cambridge University Press Cambridge, UK, 1998).
19. J. Hofbauer and K. Sigmund, Bull. Am. Math. Soc. **40**, 479 (2003).
20. P. Erdős and A. Renyi, Publicationes Mathematicae Debrecen **6**, 290 (1959).
21. A. Barabási and R. Albert, Science **286**, 509 (1999).
22. F. C. Santos and J. M. Pacheco, J. Evol. Biol. **19**, 726 (2006).
23. J. Gómez-Gardeñes, M. Campillo, L. M.Floría, and Y. Moreno, Phys. Rev. Lett. **98**, 108103 (2007).

Chapter 5
The Prisoner's Dilemma Game on Random Scale-Free Networks

As it has been well established in previous chapters, when implementing the Prisoner's Dilemma (PD) game on top of complex networks, the scale-free (SF) topologies greatly enhance cooperation [1–12], compared to other topologies as ER networks. It is also well known that the heterogeneity on the degree distribution of these structures is a crucial factor in order to achieve such high levels of cooperation in the system. More specifically, the hubs, or nodes with the highest connectivity, act always as cooperators, surrounding themselves with middle-class cooperators, and creating a unique cluster (or 'Eden') of cooperation that is able to resist the attack of defectors, even when cooperation gets really expensive. Nonetheless, up to now we have only focused on the BA model [13], among other SF network models available in literature (for a quick review of some of them, see [14, 15]). BA SF networks have some correlations by construction, the so-called age-correlations [16–18]. That means that older nodes, the ones that arrived earlier to the system when it was being built are interconnected, since they formed the original core of nodes, and besides, these older nodes usually become hubs as the network grows. The existence of age-correlations can be found in some real systems also, such as the collaboration or citation networks, or the 'old boy' network, made up of former students of the Ivy League that now work at the top investment banks [19].

In this chapter we want to study the evolution of cooperation in *random* SF structures, it is to say, those without any kind of correlations. We presume that these age-correlations among the highly connected individuals of BA networks enhance cooperation [1, 3], by making the single cooperator cluster even more robust to the possible invasion of defectors. Thus, now it is our intention to analyze the situation when considering the same PD dynamics taking place on top of a randomized version of BA topologies. Our first goal in the study of such random SF networks is to check if the deletion of the hub-to-hub links affects indeed the microscopic organization of cooperation observed in BA networks, explaining qualitatively the drop in the cooperation level as a break down of the cohesive arrangement of cooperators.

We want to study in detail the structure of cooperation in random SF networks, and in order to do so, on the one hand we will perform our usual numeric simulations.

J. Poncela Casasnovas, *Evolutionary Games in Complex Topologies*, Springer Theses,
DOI: 10.1007/978-3-642-30117-9_5, © Springer-Verlag Berlin Heidelberg 2012

Specifically, we will perform a rewiring process of the SF networks obtained by means of the BA model, which is a procedure that destroys any kind of correlations present in the original system [18], preserving the connectivity of every node, and therefore the original degree distribution, and then we will implement the usual PD dynamics. On the other hand, we will also address the problem analytically, by using a degree-based mean field approximation in order to try and incorporate the heterogeneity in the number of social contacts of individuals in the Replicator Equation [20–23] (see also Sect. 2.2.2). To this end, we will make a further compartmentalization of the strategists in degree-classes, by defining the fraction of cooperators and defectors with degree k, so we will have an equation for the evolution of the cooperation in every class of connectivity k. Finally, we will compare the results obtained with both methods, discussing whether or not this approximation is accurate enough to explain some of the basic behaviors of the cooperation in the system.

5.1 Numerical Simulations on Random Scale-Free Networks

To study the structure and dynamics of cooperation in random SF networks we have performed a rewiring process [24] of SF networks built via BA mechanism. As we have already seen in Sect. 2.1.3, the BA model makes the network grown from an initial core of m_0 nodes, incorporating a new node to the network every time step. Besides, every new node launches m links to the nodes already present in the network, following a preferential attachment rule, i.e., the probability of receiving a link from the new node is proportional to the degree of the nodes. The networks generated using the BA model have a power-law degree distribution, $P(k) \sim k^{-\gamma}$, with $\gamma = 3$. Nonetheless, they possess important features that make them different from random SF networks built by means of purely statistical algorithms such as the Molloy-Reed configurational model [25]. These differences are the previously mentioned age-correlations that have as a consequence the interconnection of highly-connected elements or hubs. The links between hubs have been shown to play a crucial role in the survival of cooperation [1, 3], since the cooperation level decreases notably when they are removed.

The rewiring process is made as follows (see Fig. 5.1): let i and j be a pair of neighbors, so they share a link, and let be m and n be another pair of nodes linked together. Then we interchange the $i-j$ and the $m-n$ links, in such a way that in the final state, $i-n$ and $m-j$ are the new pairs of neighbors. Of course, we make sure that $i \neq j \neq m \neq n$, to avoid double links and auto-links, i.e., links that connect a node with itself. We repeat the process N times, checking that the final networks have a unique connected component.As we have mentioned before, applying this rewiring scheme destroys any kind of correlations present in the original network preserving the degree sequence of the graph, and thus keeping the same degree distribution ($P(k) \sim k^{-3}$) as in the original BA network.

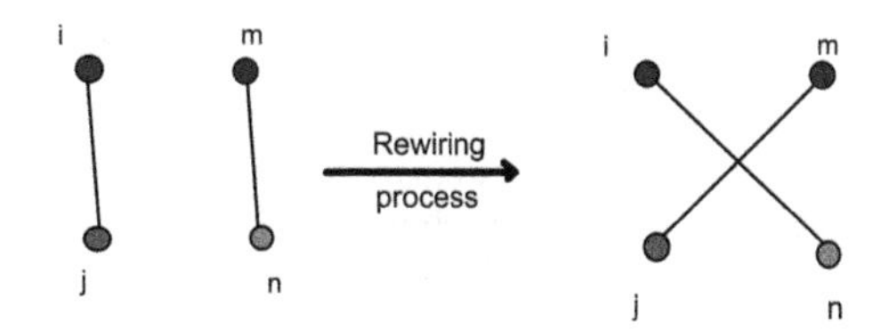

Fig. 5.1 Schematic representation of the rewiring process of two pairs of nodes

Once the network is rewired, we perform the numerical simulation of the evolutionary dynamics dictated by the Prisoner's Dilemma, whose payoff matrix is given, as usual, by:

$$\begin{array}{cc} & \begin{array}{cc} C & D \end{array} \\ \begin{array}{c} C \\ D \end{array} & \begin{pmatrix} R & S \\ T & P \end{pmatrix} \end{array} \tag{5.1}$$

where we set, again $P = S = 0$, $R = 1$, $T = b > 1$, so we only have to deal with one control parameter, the temptation to defect b [26, 27, 1].

In the initial configuration of the system, the probabilities of being a cooperator or a defector are the same ($\rho_0 = 0.5$), and the strategists are *randomly distributed* across the network. On the other hand, we will use the same updating rule as in previous chapters, the Replicator-like rule [1, 2, 22, 28–30], so player i adopts the strategy of its neighbor j for the next game round with probability:

$$\Pi_{i\to j} = \beta(P_j - P_i) \tag{5.2}$$

where P_i and P_j are their correspondent payoffs from the last round of the game, and with $\beta = (\max\{k_i, k_j\}b)^{-1}$.

The details of the numerical simulations are similar to those in previous chapters: the networks we generated have $N = 4 \cdot 10^3$ nodes and an average connectivity $\langle k \rangle = 4$. We let the system evolve until a stationary regime is reached. This stationary regime is characterized by an average level of cooperation that is the fraction of C players in the network, $\langle c \rangle = c/N$. To compute $\langle c \rangle$ we let the dynamics evolve over a transient time $\tau_0 = 5 \cdot 10^3$, and we further let the system evolve over time windows of $\tau = 10^3$ generations. In each time window, we compute the average value and the fluctuations of $c(t)$. When the fluctuations are less than or equal to $1/\sqrt{N}$, we stop the simulation and consider the average cooperation obtained in the last time window as the asymptotic average cooperation $\langle c \rangle$. In order to make an extensive sampling of initial conditions and network realizations we have performed 10^3 independent numerical simulations for each value of the temptation to defect b studied, and averaged the values $\langle c \rangle$ found in the realizations.

First of all, in Fig. 5.2 we show a comparison of the levels of cooperation achieved by such random SF networks, as well as original BA and ER topologies, and as it can be seen, our results confirm previous findings: the removal of age-correlations makes random SF networks much less robust to defection than BA networks [2, 3], so the level of cooperation drops substantially. On the other hand, in Fig. 5.3a we

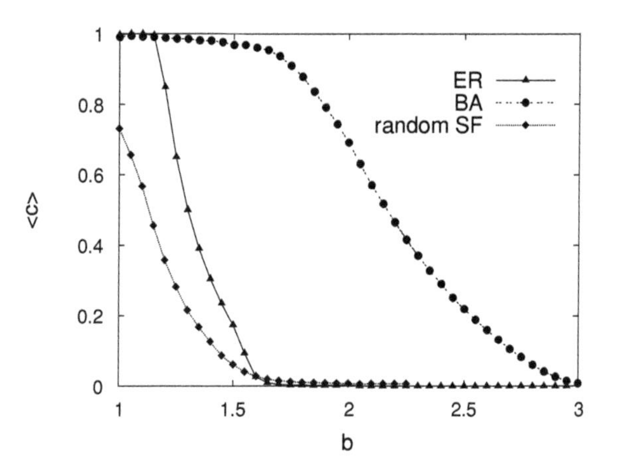

Fig. 5.2 Comparison of the levels of cooperation achieved in the stationary state for ER, BA and random SF networks, as a function of the temptation to defect b. All networks are made up of $N = 4 \cdot 10^3$ nodes and have an average connectivity $\langle k \rangle = 4$. Every point shown is the average over 10^3 different realizations

have also plotted the average level of cooperation $\langle c \rangle$ as a function of b, along with the level of pure strategists and fluctuating individuals present on the network. It is to say, on these topologies we have also found that there is a partition of the network into pure strategists (pure cooperators PC and pure defectors PD), and fluctuating individuals (F) on the stationary regime. Notice that the partition of the system into pure strategists and fluctuating individuals has been made following the same criteria as in Sect. 3.3. As one could expect, the fraction PC decreases with b, the fluctuating take over the network for a wide range of intermediate values of b, and the PD finally invade the system for high values of the parameter. Nonetheless, the fraction of PC is remarkably lower than that for the case of BA networks or even ER topologies, whereas the fluctuating individuals dominate the system for a wider range of b, so the level of cooperation is almost exclusively due to them. This is a very different scenario from those studied for BA SF networks (compare with Fig. 3.2).

Moreover, in Fig. 5.3b we have plotted the number of cooperator clusters N_{cc} and defector clusters N_{dc} as a function of b, using to that aim the same definition as in Sect. 3.7: a cooperator (defector) cluster is a connected subgraph composed of nodes that are pure cooperators (defectors). The first difference with respect to BA networks is that here we find realizations with more than one cooperator cluster, whereas for BA networks, the number of clusters was always exactly $N_{cc} = 1$, as long as $\langle c \rangle(b) > 0$. This difference explains the drop in the cooperation level previously observed [1]: the more fragmented the cooperators are arranged, the less sources of benefits they find in their surroundings and the larger is the probability to be invaded by the instantaneous defectors that are in contact with them. Regarding the defector clusters we observe the same picture as in BA networks: PD are arranged in several clusters when they start to invade the network ($b \gtrsim 2$). The number of defector clusters decreases as they start to grow in size and glue together, and finally collapse into a single one, when all the network has been totally invaded by pure defectors.

We have also checked the probability that a node of degree k is a cooperator in the stationary regime. Our numerical simulations show that high degree nodes are more

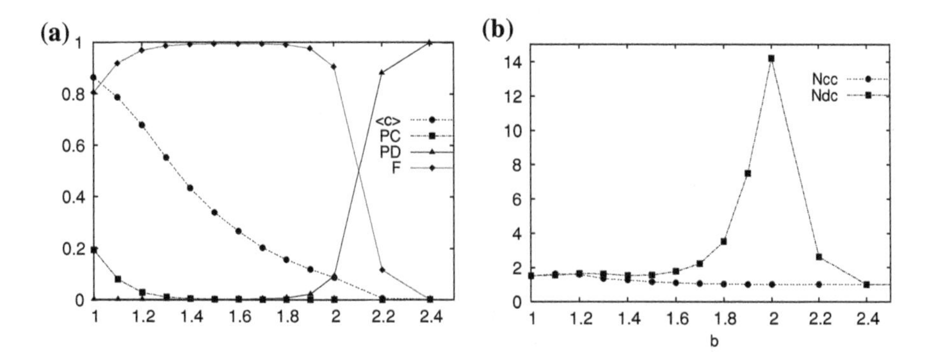

Fig. 5.3 **a** Average level of cooperation $\langle c\rangle$ as a function of the temptation to defect b in random SF graphs. The panel also shows the corresponding dependence of the fraction of pure cooperators (PC), pure defectors (PD) and fluctuating (F) players. **b** Average number of cooperator clusters N_{cc} and defector clusters N_{dc} as a function of b. The networks are made up of $N = 4 \cdot 10^3$ nodes and an average connectivity $\langle k\rangle = 4$. Every point shown is the average over 10^3 different realizations

likely to act as cooperators than intermediate or low degree individuals, in agreement with previous numerical observations we have made in BA networks (see Sect. 3.5).

Summing up, in random SF networks the fragmentation of the cooperator clusters together with the extremely low fraction of pure cooperators and the prevalence of fluctuating individuals not only makes the average level of cooperation drop in comparison with that same PD dynamics on top of BA networks, but also lead to an organization of cooperation that is quite different to that observed in BA SF networks. Therefore, we can confirm that the high level of cooperation that BA SF networks can hold is not only due to its degree distribution, buy also due to the so-called age-correlations that link together the hubs.

5.2 The Degree-Based Mean Field Approximation

The random SF graphs used in the simulations above are free of any kind of correlation between the properties of two adjacent nodes (age, degree, etc...). Therefore, it is reasonable to try and study analytically the evolution of cooperation in these systems by considering a similar approach to that used for disease spreading processes in complex networks with arbitrary degree distributions and no correlations [31–33]. To incorporate the heterogeneity in the number of social contacts of individuals we make a further compartmentalization of the strategists in degree-classes. In this sense, we label c_k and d_k the fractions of cooperators and defectors with degree k, respectively, so that the total number of cooperators and defectors will be:

$$c = N\sum_k P(k)c_k \ , \tag{5.3}$$

$$d = N \sum_{k} P(k) d_k \, . \tag{5.4}$$

Obviously the relation $c_k + d_k = 1$ holds, and, instead of describing the evolution of the fraction of cooperators in the population via the well-known Replicator Equation [20–23], we can write now the evolution of the fraction of cooperators with degree k as:

$$\dot{c_k} = (1 - c_k)\Pi_k^{DC} - c_k \Pi_k^{CD} \, , \tag{5.5}$$

where Π_k^{DC} is the probability that a defector of degree k changes its strategy to cooperation, and analogously, Π_k^{CD} is the probability that a cooperator of degree k change its strategy to defection.

Assuming that the network has no degree–degree correlations, and following the replicator-like update rule (5.2), we can write the probabilities Π_k^{DC} and Π_k^{CD} as:

$$\Pi_k^{DC} = \sum_{k'} \frac{k' P(k')}{\langle k \rangle} \beta \, \Theta\left[P_{k'}^C - P_k^D\right] c_{k'} \, , \tag{5.6}$$

$$\Pi_k^{CD} = \sum_{k'} \frac{k' P(k')}{\langle k \rangle} \beta \, \Theta\left[P_{k'}^D - P_k^C\right] (1 - c_{k'}) \, , \tag{5.7}$$

where the function $\Theta[x]$ is defined as $\Theta[x] = x$ if $x > 0$ and $\Theta[x] = 0$ otherwise. Besides, P_k^C and P_k^D are the payoffs obtained by a cooperator and a defector of degree k respectively, and can be written as:

$$P_k^C = k \sum_{k'} \frac{k' P(k')}{\langle k \rangle} c_{k'} = k l_c \, , \tag{5.8}$$

$$P_k^D = b \cdot k l_c \, , \tag{5.9}$$

where l_c is the probability that a node has a cooperator neighbor. Now we can insert the above two expressions (5.8) and (5.9) in equations 5.7 and 5.6 and finally write the evolution equation of the fraction of cooperators with degree k (5.5) as

$$\begin{aligned} \dot{c_k} = & (1 - c_k) \sum_{k' > bk} \frac{k' P(k')}{\langle k \rangle} \beta \, l_c (k' - bk) c_{k'} \\ & - c_k \sum_{k' > bk} \frac{k' P(k')}{\langle k \rangle} \beta \, l_c (bk' - k)(1 - c_{k'}) \\ & - c_k \sum_{k' > k/b}^{bk} \frac{k' P(k')}{\langle k \rangle} \beta \, l_c (bk' - k)(1 - c_{k'}) \, , \end{aligned} \tag{5.10}$$

where we have separated the contributions to the transition C→D that come from neighbors with $k' > bk$ and $k' < bk$, so that it is clear that the number of degree classes that participate in the transition C→D is larger than those that influence the change D→C.

We have numerically solved the set of equations 5.10 using both power-law and a Poisson distribution for the generic expression of the degree distribution $P(k)$. As initial conditions, we have used a homogeneous distribution of cooperators and defectors for all the degree classes: $c_k(t = 0) = a \; \forall k$ where a is a random variable homogeneously distributed between [0, 1]. This way, the initial fraction of cooperation is $\rho_0 = 0.5$, in agreement with the numerical experiments shown in the previous sections.

Unfortunately, the numerics clearly showed that the total cooperation always decays to zero whenever $b > 1$, thus failing to explain the cooperation levels observed in the numerical simulations in both random SF networks and ER graphs. Nonetheless, this result is consistent with previous findings, which have shown that the mean field approximation can not explain satisfactory the observed survival of cooperation. However, in the next section we will study the behavior of the system when it starts from a very specific set of initial conditions: *targeted cooperation*.

5.3 Targeted Cooperation

We have failed to use the degree-based mean field approximation to explain the observed non-zero level of cooperation when simulating the PD dynamics on top of random SF networks. Now we study a very particular case for both random SF network simulations and our degree-based mean field approximation with a particular set of initial conditions. As we will see next, the results show that at least, if not in perfect agreement, the two cases show similarities on the qualitative behavior of both the time evolution $C(t)$ towards the stationary state and the final state achieved by the state, expressed through the dependence $\langle c \rangle (b)$.

It is important to stress that the main assumption behind the above mean field approach is that the average level of cooperation inside a degree-class, c_k, is a *properly defined magnitude* for describing the state of the nodes within that degree. In particular, this assumption is strictly correct when c_k is either 1 or 0. This motivated us to study the solution of Eq. 5.10 using a particular set of initial conditions that we have called the *targeted cooperation*, and that are explained next.

We define targeted cooperation as a set of initial conditions for the system described by 5.10, where $c_k(t = 0) = 1$ if $k > k^*$ and $c_k(t = 0) = 0$ if $k < k^*$. It is to say, all nodes whose connectivity is higher than a given value k^* are set initially as cooperators, while all those with lower number of neighbors will be defectors. We have carefully explored the solutions of Eq. 5.10 when $P(k)$ is a power-law degree distribution. To this end, we have considered power-law distributions with several values of the exponent γ, and we have also used different values for the degree threshold k^*. The numerical solution of Eq. 5.10 reveals that, in this case, the cooperation

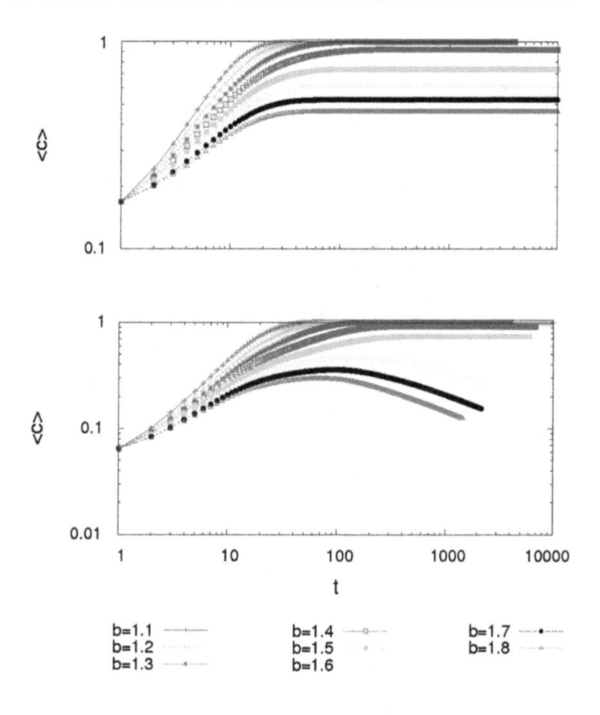

Fig. 5.4 Time evolution of the fraction of cooperators $\langle c\rangle(t)$ obtained solving equation 5.10 numerically, when targeted cooperation is used as initial conditions and being $P(k)$ a power-law with $\gamma = 3$. The different curves correspond to several values of b, as shown in the bottom of the figure. The targeted cooperation used correspond to (**a**) $k^* = 2$ and (**b**) $k^* = 3$. Notice the log–log representation of the axes

survives for $b > 1$, reaching a stationary value that depends on both the value of b and that of the threshold k^*. In figure 5.4 we show the time evolution of the average level of cooperation for several values of b and $k^* = 2$ and $k^* = 3$. The degree distribution in the figure is a power-law with $\gamma = 3$. The solutions show that the larger k^* and/or b are, the lower the cooperation level is, which makes perfect sense, since they imply, respectively that the number of initial cooperators is lower, or that the cooperation itself gets more expensive.

On the other hand, it is interesting to compare these results with the values obtained for our conventional simulations on top of random SF networks (see Fig. 5.5). We see that the behavior of both systems are relatively similar, as far as time evolution of the cooperation is concerned (but, of course, the evolution of the random SF networks displays finite size fluctuations). As it can be seen in the (Left) panel of Fig. 5.5, for a fixed value of k^* and for low or medium values of b, the level of cooperation increases with time, until it gets its final value (which depends inversely on b), and for higher values of b, the level of cooperation on the system eventually goes to zero. Conversely, if we fix the value of b ((Right) panel of Fig. 5.5), the higher the k^*, the lower the final level of cooperation the system can achieve. Besides, in Fig. 5.6 we show the dependence of the level of cooperation $\langle c\rangle$ with both the temptation to defect b and with the value of the threshold k^*.

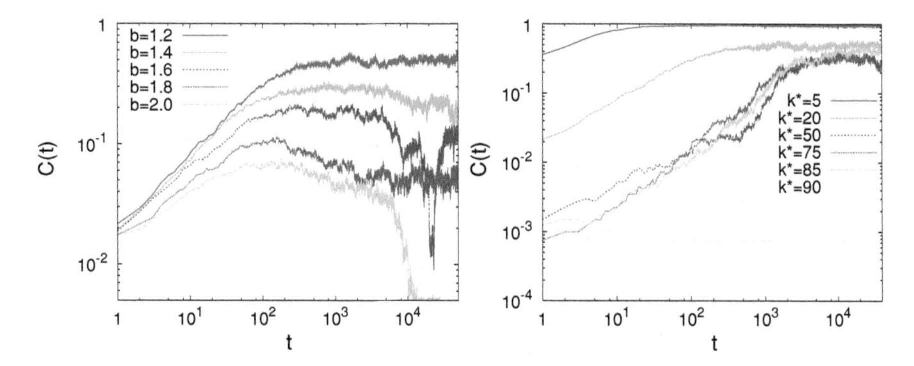

Fig. 5.5 Several examples of individual time evolutions of the random SF network simulations for targeted cooperation. All the cases shown in the *Left* panel have $k^* = 20$, while those in the *Right* one, correspond to simulations with a fixed value of $b = 1.2$. The networks are made of $N = 4 \cdot 10^3$ nodes, with average connectivity $\langle k \rangle = 4$ and $\gamma = 3$. Notice the log-log representation of the axes

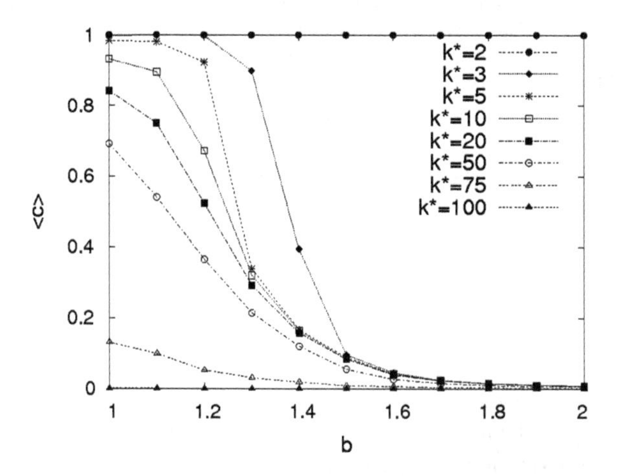

Fig. 5.6 The average level of cooperation $\langle c \rangle$ as a function of the temptation to defect b, for several values of the degree threshold k^* for simulations on top of random SF networks with targeted cooperation. The networks are made of $N = 4 \cdot 10^3$ nodes, with average connectivity $\langle k \rangle = 4$ and $\gamma = 3$

5.4 Dependence with the Exponent of the Power-Law Distributions for the Mean Field Approximation

Returning now to the degree-based mean field approach, it is interesting to study in detail the effect of the degree threshold k^* over the asymptotic level of cooperation. In particular, we can focus on the minimum amount of degree classes that we have to fill initially with cooperators so that cooperation is able to survive asymptotically in the system. We have carefully explored different sets of initial conditions corresponding to different values of k^*. Starting from a low value of k^* we have solved Eq. 5.10

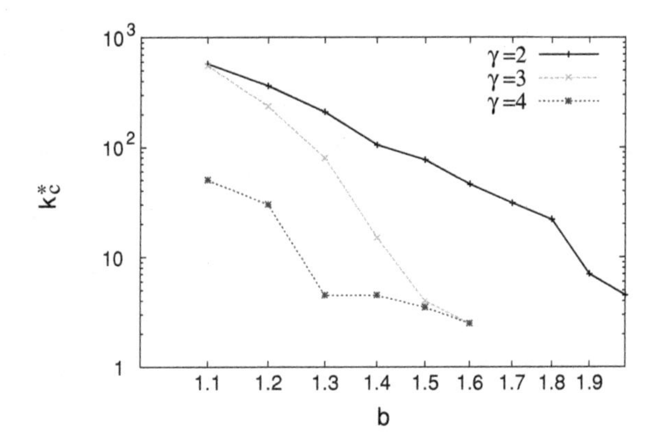

Fig. 5.7 Phase diagram $k_c^*(b)$. The three curves correspond to different power-law distributions (namely, $\gamma = 4$, 3 and 2). Each curve $k_c^*(b)$ represent the border between two different asymptotic regimes for the evolution of Eq. 5.10 with targeted cooperation: The area below the curves correspond to the points (b, k^*) where targeted cooperation yield nonzero asymptotic level of cooperation. Conversely, the area above the curves correspond to the targeted initial conditions for which the evolution of Eq. 5.10 yields $\langle c \rangle \to 0$

and computed the final level of cooperation $\langle c \rangle$. If $\langle c \rangle > 0$ we increase the value of k^* and solve again the system 5.10. This process is iterated until we reach a value k_c^* for which cooperation finally vanishes. The critical value k_c^* represents the minimal amount of cooperator degree classes needed at time 0 to sustain asymptotically a nonzero level of cooperation. In Fig. 5.7 we have plotted the functions $k_c^*(b)$ for three power-law degree distributions ($\gamma = 2$, 3 and 4). Obviously, we observe that as the cooperation gets more and more expensive, it is necessary to fill more degree classes to assure a nonzero level of cooperation. More interestingly, we show that the heterogeneity of the network (or in other words, a lower value for the exponent γ in the degree distribution $P(k)$) increases the value of k_c^*. This result is related to the fact that filling a given amount of degree classes is more efficient (more nodes are initially set as cooperators) when the network is more heterogeneous.

5.5 Comparison Between Simulations and Mean-Field Approximation for the Targeted Cooperation Initial Conditions

We can say that the mean field approach represents a useful tool for substituting computationally expensive numerical simulations to a given extent. However, how accurate are the results of the solutions of Eq. 5.10 when compared to simulations on top of networks with targeted cooperation as the initial condition? To check the reliability of the degree-based mean field approach in the context of targeted

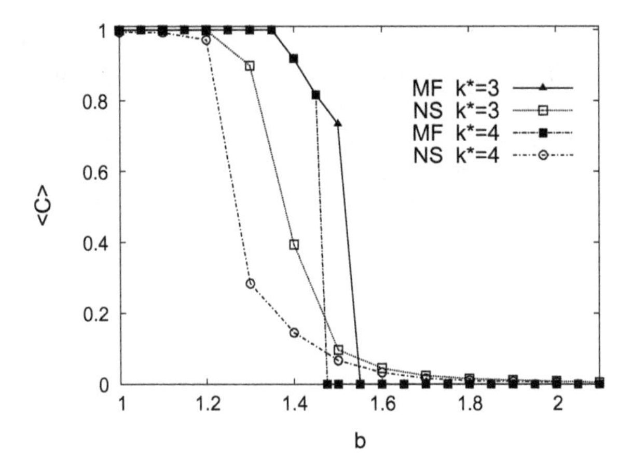

Fig. 5.8 Evolution of the asymptotic level of cooperation $\langle c \rangle$ obtained when (*i*) solving the mean field (MF) Eq. 5.10 and (*ii*) computed through numerical simulations (NS) of the evolutionary dynamics on top of a random SF network. The degree distribution used is a power-law with $\gamma = 3$. In both cases we have set targeted cooperation as initial conditions for the evolutionary dynamics. We have used $k^* = 3$ and 4

cooperation we have computed the diagram $\langle c \rangle(b)$ for random SF networks with $\gamma = 3$ using two different sets of initial conditions corresponding to $k^* = 3$ and 4. In Fig. 5.8 we show the results of the simulations compared to the results obtained by solving Eq. 5.10. Obviously, the agreement is not complete but we can say that the dependence of the level of cooperation with the temptation to defect b follows the same qualitative behavior and the cooperation tends to zero around the same values of b.

The values of b for which $\langle c \rangle = 0$ in each of the curves of the figures are obviously related to the values k_c^*. Our results show that, although the level of cooperation starts decreasing earlier (it is to say, for lower values of b), the curves $\langle c \rangle(b)$ obtained from simulations on top of random SF networks can hold larger values of b with $\langle c \rangle > 0$ than the system described by Eq. 5.10. On the other hand, the simulations yield very low (but yet non-zero) values of $\langle c \rangle$ for those values of b for which cooperation asymptotically vanishes solving Eq. 5.10. The drop of the level of cooperation is much more abrupt for the mean field scenario. Therefore, this mean field approach seems to be, at least, of help to study the behavior of $k_c^*(b)$ and the asymptotic level of cooperation of the system when targeted cooperation is initially placed in the system.

Regarding general (i.e., non-targeted cooperation type of) initial conditions for the degree-based mean field Eq. 5.10, some comments are in order. For both, power-law and Poisson degree distributions $P(k)$, random uniformly distributed values for $c_k(t = 0)$, as well as fixed value $c_k(t = 0) = 0.5$ (mimicking the initial conditions in the numerical simulations of previous section), led to asymptotic zero level of cooperation as soon as $b > 1$. This suggests that, generically speaking, mean field approaches to the evolutionary dynamics of prisoner's dilemma games on graphs (even in generalized forms, as Eq. 5.10) are likely bound to fail to account for the numerically observed survival of cooperation. This would be in agreement with some previous results on a particular type of artificial networks that allow a rigorous analysis of the issue [34]. To put it in plain terms, the network reciprocity mechanisms that enhance the evolutionary survival of cooperation in network settings [35] seem to be out of reach of the (homogeneity) mean field assumptions, in the sense that they are associated in an essential way to fluctuations of averaged quantities, like c_k which are the basic descriptors in mean field approaches. Besides, the existence of loops and cycles is also a mechanism able to promote cooperation that is overlooked by the mean field approach.

5.6 Conclusions

Scale-free networks have been recently shown as the graphs that better promote cooperation. In this chapter we have shown that the power-law degree distribution cannot be considered as the only root for the promotion of cooperation. At variance with the BA networks, the SF graphs considered in this chapter are free of any kind of node-node correlation. The first conclusion of our study is that we confirm the previous finding pointing out the fact that cooperation decays when no correlations are present in the network. Moreover, we have shown that the organization of cooperation is dramatically different from that of the BA network, showing that cooperators can arrange in more than one cluster, increasing the probability of being invaded by defectors. In other words, the fixation of cooperation is much lower than in SF networks with correlations, thus completing the picture provided by other studies where correlations were added into SF networks enhancing the promotion of cooperation of BA networks [36, 37]. On the one hand, our study in random SF networks can be considered as the null model for the study of the cooperation in other types of SF graphs. Besides, our results highlight the importance of taking into account other structural properties beyond the degree distribution of the network [38] in order to capture the mechanisms that help to fixate cooperation in real complex networks.

The second part of the chapter presents a degree-based mean field approach to analytically study networks with an arbitrary degree distribution and no node-node correlations (such as random SF networks). The approach relies on a degree compartmentalization of cooperators and defectors strategists. We have shown that, contrary to diffusion dynamics where a similar approach has been applied successfuly [31–33], the degree-based mean field equations do not work correctly when general initial conditions are applied, since no asymptotic level of cooperation is observed when the temptation to defect is larger than the reward for cooperation ($b > R = 1$). On the other hand, when a particular set of initial conditions is used (consisting in placing all the cooperators in the higher degree classes of the network) the solution of the mean field yields a non zero level of cooperation for a number of targeted initial configurations. The results obtained in this latter context qualitatively agree with those obtained when extensive simulations on top of random SF graphs are performed.

As a conclusion, the results presented in this chapter complete the studies about the Prisoner's Dilemma on top of SF networks showing that node-node correlations play a key role for sustaining a high level of cooperation. In this line, the wrong functioning of the degree-based mean field approach further confirms that heterogeneity is not the unique responsible of enhancing cooperation. The presence of features that are beyond the scope of this mean field formulation (even in uncorrelated graphs) such as cycles or loops seems to be at the root of cooperation enhancement.

References

1. F. C. Santos and J. M. Pacheco, Phys. Rev. Lett. **95**, 098104 (2005).
2. F. C. Santos, F. J. Rodrigues, and J. M. Pacheco, Proc. Biol. Sci. **273**, 51 (2006).
3. F. C. Santos and J. M. Pacheco, J. Evol. Biol. **19**, 726 (2006).
4. F. C. Santos, J. M. Pacheco, and T. Lenaerts, Proc. Natl. Acad. Sci. USA **103**, 3490 (2006).
5. H. Ohtsuki, E. L. C. Hauert, and M. A. Nowak, Nature **441**, 502 (2006).
6. G. Abramson and M. Kuperman, Phys. Rev. E **63**, 030901(R) (2001).
7. V. M. Eguíluz, M. G. Zimmermann, C. J. Cela-Conde, and M. San Miguel, American Journal of Sociology **110**, 977 (2005).
8. T. Killingback and M. Doebeli, Proc. R. Soc. Lond. **263**, 1135 (1996).
9. G. Szabó and G. Fáith, Phys. Rep. 446, 97 (2007).
10. A. Szolnoki, M. Perc, and Z. Danku, Physica A **387**, 2075 (2008).
11. J. Vukov and G. S. A. Szolnoki, Phys. Rev. E **77**, 026109 (2008).
12. J. Gómez-Gardeñes, M. Campillo, L. M. Floría, and Y. Moreno, Phys. Rev. Lett. **98**, 108103 (2007).
13. A. Barabási and R. Albert, Science 286, 509 (1999).
14. S. Boccaletti, V. Latora, Y. Moreno, M. Chavez, and D. Hwang, Phys. Rep. **424**, 175 (2006).
15. G. Caldarelli, A. Capocci, P. D. L. Rios, and M. A. M. noz, Phys. Rev. Lett. **89**, 258702 (2002).
16. S. N. Dorogovtsev and J. F. F. Mendes, *Evolution of networks. From biological nets to the Internet and the WWW.* (Oxford University Press, Oxford, UK, 2003).
17. M. Newman, SIAM Review **45**, 167 (2003).
18. R. Albert and A. L. BarabÃ¡si, Rev. Mod. Phys. **74**, 47 (2002).
19. S. H. Strogatz, Nature **410**, 268 (2001).
20. J. Hofbauer, P. Schuster, and K. Sigmund, J. Theor. Biol. **81**, 609 (1979).
21. P. Taylor and L. Jonker, Math. Biosci. **40**, 145 (1978).
22. H. Gintis, *Game theory evolving*. (Princeton University Press, Princeton, NJ, 2000).
23. H. Ohtsuki and M. A. Nowak, J. Theor. Biol. *243*, 86 (2006).
24. S. Maslov and K. Sneppen, Science **296**, 910 (2002).
25. M. Molloy and B. Reed, Combinatorics, Probability and Computing **7**, 295 (1998).
26. K. Lindgren and M. Nordahl, Physica D **75**, 292 (1994).
27. M. A. Nowak and R. M. May, Nature **359**, 826 (1992).
28. C. Hauert and M. Doebeli, Nature **428**, 643 (2004).
29. J. Hofbauer and K. Sigmund, *Evolutionary games and population dynamics*. (Cambridge University Press, Cambridge, UK, 1998).
30. J. Hofbauer and K. Sigmund, Bull. Am. Math. Soc. *40*, 479 (2003).
31. R. Pastor-Satorras and A. Vespignani., Phys. Rev. Lett. **86**, 3200 (2001).
32. R. Pastor-Satorras and A. Vespignani., Phys. Rev. E *63*, 066117 (2001).
33. Y. Moreno, R. Pastor-Satorras, and A. Vespignani., European Physical Journal B **26**, 521 (2002).

34. L. M. Floría, C. Gracia-Lá¡zaro, J. Gómez-Gardeñes, and Y. Moreno, Phys. Rev. E **79**, 026106 (2009).
35. M. Nowak, Science **314**, 1560 (2006).
36. S. Assenza, J. Gómez-Gardeñes, and V. Latora, Phys. Rev. E **78**, 017101 (2008).
37. A. Pusch, S. Weber, and M. Porto, Phys. Rev. E **77**, 036120 (2008).
38. L. Costa, F. A. Rodrigues, G. Travieso, and P. R. V. Boas., Advances in Physics **56**, 167 (2007).

Chapter 6
The Prisoner's Dilemma Game on Scale-Free Networks with Limited Number of Interactions

It has been widely studied in the literature how on complex networks, far from the well-mixed assumption or regular lattices [1], cooperation has much better chances to survive, even when it gets very expensive [2–5]. Specifically, it has been proved that heterogeneity not only reproduces much better some topological features of the social systems [6, 7], such as the degree distribution, but also greatly favors cooperation. This happens, as we have seen in some detail in Chap. 3, thanks to the formation of one single cluster, centered on the interconnected cooperator hubs, that create a 'supporting system' for the individuals, in order to resist invasions from defectors [8, 9]. Nonetheless, when modeling some aspects of the behavior of individuals in a society using evolutionary games on complex networks, usually the number of interactions a node establishes in every round is considered equal to the number of topological neighbors it has. This widely used assumption does not take into account real constrains such as the limited amount of time to deal with social acquaintances nor the energy it costs to the node to pay attention to each of its neighbors.

In this chapter we analyze a more realistic scenario in which agents are limited to interact with a given number of neighbors during each round of the game. In particular, we will study the effect of such a restriction in the number of interactions per round of the evolutionary Prisoner's Dilemma game on scale-free networks. In this sense, some effort has been put on studying the effect of restricting the maximum number of possible contacts a node can have due to the finite resources of the node, but in a different way than the approach we propose now. In [10], the level of cooperation achieved by the system is studied when the SF networks have a cutoff at a certain value for the connectivity, k_{cutoff}, so there will be no nodes with a number of connections above that given value. In this scenario, it was found that the level of cooperation remains high enough even for an important cutoff of the degree distribution (up to a value $k_{cutoff} > 20$ for a network made up of $N = 10^4$ nodes), and what is more, some slight improvement can be found in the average cooperation as the value k_{cutoff} decreases, as long as it is larger than a certain threshold $k_{cutoff} \lesssim 20$.

It is also worth mentioning that, a different approach but in the same direction of restricting somehow the available resources for a node has been used when dealing

J. Poncela Casasnovas, *Evolutionary Games in Complex Topologies*, Springer Theses,
DOI: 10.1007/978-3-642-30117-9_6, © Springer-Verlag Berlin Heidelberg 2012

with the Public Goods Game. In [11], Santos et. al. compared the level of cooperation in the system for two scenarios: a fixed-cost-per-individual situation where a node with connectivity k contributes $c/(k+1)$ in every one of the $(k+1)$ rounds of the game, and a fixed-cost-per-interaction where it contributes c in every round of the game, regardless of its connectivity. They found that the former situation promotes cooperation more than the latter, due to the introduction of an extra source of heterogeneity, apart from the topological one. Namely, this diversity in the amount that every node contributes to the common goods has been proved to be beneficial for the overall level of cooperation in the system.

Nonetheless, we want to address this restriction from a different angle: the degree distribution of the topological substrate remains untouched, it is to say, the PD dynamics will take place on top of unaltered BA scale-free networks. However, every node i of the network, even when it has k_i topological connections, will only be allowed to establish k^* interactions per round of the game among its neighbors. This restriction is the same for all nodes in the system, but it will specially affect those nodes having a large topological connectivity, the hubs, that will only play with a small fraction of their otherwise large number of neighbors, while it will not affect those nodes with a very low connectivity at all. We will analyze the consequences that limiting the number of game mates may have on the global dynamics of the system, and more precisely on the average level of cooperation, comparing the results with the well-known case of a standard framework in which every node plays every round of the game with all its neighbors, as dictates the underlying topology.

One should also keep in mind that the formulation of the Prisoner's Dilemma that will be used in this chapter is different from the one used in previous chapters. It means that the specific values of the coefficients of the payoff matrix will be different, but not their relative ordering. In this way, now, instead of having the temptation to defect, b as the (only) free parameter, we will have the ratio b/c, between the benefit of playing against a cooperator and the cost of being one. This particular formulation will be used again in Chap. 8.

6.1 The Model

We use scale-free networks built via the Barabási-Albert (BA) preferential attachment model [12]. As we have already explained (see Sect. 2.1.3), the well-known BA model is based on growth and preferential attachment, and starting from a small set of m_0 fully connected nodes, every time step we add a new node j to the network. This new node will attach to m of the existing nodes. The probability that a link from node j connects to an existing node i is proportional to its degree, $P_i = \frac{k_i}{\sum_l k_l}$. This procedure continues until the network reaches its final size N. The degree distribution of such networks is a power-law, $P(k) \sim k^{-\gamma}$, with an exponent $\gamma = 3$ and the average connectivity is $\langle k \rangle = 2m$. In our case, we have used networks with $N = 4 \times 10^3$ nodes and an average value for the connectivity $\langle k \rangle = 4$.

We consider that every node on the network is a player whose initial strategy, cooperator (C) or defector (D), is randomly assigned with equal probability $\rho_0 = 0.5$. Next, we go node by node, forcing them to choose, also randomly, k^* among its k_i topological neighbors, so we get an 'effective connectivity matrix' for the current round of the game. Obviously, if $k_i \leq k^*$ for a particular node i, then it chooses all its neighbors to play with them every single time, but if $k_i > k^*$, then it will play only with some of them, making a different selection every round. Notice that, in order to preserve the symmetry of the interactions, if node i chooses node j, it means that j also chooses i straightaway (apart from those corresponding k^* neighbors that j has chosen or will choose when its time comes), so the real *effective connectivity* of the nodes is not strictly k^*, but it is in general $k_i^{eff} \gtrsim k^*$.

We can calculate the dependence of the effective connectivity k_i^{eff} with the topological connectivity of a node k_i. To this aim we distinguish between those nodes having $k_i \leq k^*$ and those with $k_i > k^*$. For the former group we trivially have $k_i^{eff} = k_i$ while for the second set we have $k_i^{eff} = k^* + k_i^{in}$. In this latter case k_i^{in} stands for the number of extra connections a node i gets from being selected by other neighbors not contained in its own set of k^* selected neighbors. We can write the expression for the extra k_i^{in} game mates as:

$$k_i^{in} = k_i \left[\sum_{k' \leqslant k^*} P(k'|k_i) + \sum_{k' > k^*} \frac{k^*}{k'} P(k'|k_i) \right] , \tag{6.1}$$

where $P(k'|k)$ is the conditional probability that a node of degree k is connected with a node of degree k'. Assuming that the network is uncorrelated (as the BA network) we have $P(k'|k) = k'P(k')/\langle k \rangle$. Taking the continuous approximation for the degree we can write Eq. 6.1 as

$$k_i^{in} \approx \frac{k_i}{\langle k \rangle} \left[\int_{k_0}^{k^*} k'P(k')dk' + k^* \int_{k^*}^{\infty} P(k')dk' \right] , \tag{6.2}$$

where k_0 is the minimum degree of the network.

Solving the right hand side of the above equation for a scale-free network, $P(k) = (\gamma - 1)k_0^{\gamma-1}k^{-\gamma}$, we obtain:

$$k_i^{in} \approx \frac{k_i(\gamma - 1)k_0^{\gamma-1}}{\langle k \rangle} \left[\frac{k_0^{2-\gamma} - (k^*)^{2-\gamma}}{\gamma - 2} + \frac{(k^*)^{2-\gamma}}{\gamma - 1} \right] . \tag{6.3}$$

In our particular case we have networks with $\gamma = 3$, $\langle k \rangle = 4$ and $k_0 = 2$, therefore the effective connectivity for those nodes with $k_i > k^*$ reads

$$k_i^{eff} \approx k^* + k_i(1 - \frac{1}{k^*}) . \tag{6.4}$$

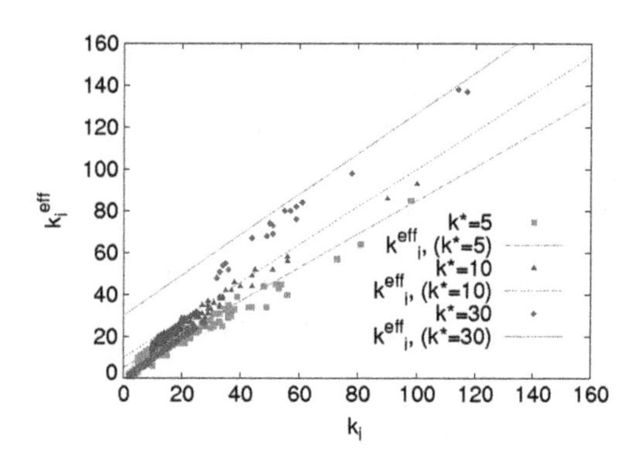

Fig. 6.1 Comparison of the actual topological connectivity of the nodes, k_i, and their effective connectivity, k_i^{eff}, and the approximate expression, for three fixed values of $k^* = 5$, $k^* = 10$ and $k^* = 30$ (a single realization of the network per each). The lines are for the theoretical estimation (Eq. 6.4)

In order to check the above approximation, we plot in Fig. 6.1 the function $k_i^{eff}(k_i)$ along with the pairs of values (k_i, k_i^{eff}) obtained in a single realization of the network when $k^* = 5$, $k^* = 10$ and $k^* = 30$, respectively. From the figure it becomes clear that the agreement with Eq. 6.4 is good.

Once all the nodes have selected their current effective neighborhood, k_i^{eff}, they play a round of the PD game with every single one of them, and accumulate their corresponding benefits π_i, according to the payoff matrix of the Prisoner's Dilemma game we are using [3, 13, 14], given by:

$$\begin{array}{c} \\ C \\ D \end{array} \begin{array}{c} \begin{array}{cc} C & D \end{array} \\ \begin{pmatrix} b-c & -c \\ b & 0 \end{pmatrix} \end{array} \sim \begin{array}{c} \\ C \\ D \end{array} \begin{array}{c} \begin{array}{cc} C & D \end{array} \\ \begin{pmatrix} b/c-1 & -1 \\ b/c & 0 \end{pmatrix} \end{array} \tag{6.5}$$

where c is the cost of being a cooperator, and b is the benefit of playing against one (obviously, the larger the ratio b/c gets, the cheaper it becomes to be a cooperator). Immediately afterwards, and in order to update its strategy, every node i compares its own payoff π_i with the payoff of one of its neighbors, π_j, randomly chosen from the current effective neighborhood. For the probability that i imitates j's strategy for the next round of the game, and following previous works [15–19], we have chosen the so-called Fermi function from Statistical Physics, given by:

$$P_{i \to j} = \frac{1}{1 + e^{w(\pi_i - \pi_j)}}, \tag{6.6}$$

where w is a parameter that accounts for the importance of the relative difference of payoffs on the change of strategy of node i. Notice that, for $w \to \infty$, the probability $P_{i \to j}$ strongly depends on the difference of payoff between the two nodes involved, so with a very high probability, if $\pi_i < \pi_j$, i will imitate j, and if $\pi_i > \pi_j$, i will not imitate j. But on the other hand, when $w \to 0$, one gets that the probability of changing strategies is $P_{i \to j} = 1/2$, independently of the values of the payoffs (in this case we have the so-called random drift evolution of the system). We can also interpret this situation as a total loss of information: the individuals know nothing at all about their neighbors, so they decide by tossing a coin [17]. The results shown on this work correspond only to the value $w = 1$. Nonetheless, we have checked that they are quite robust: when testing out other values for w we get qualitatively the same outcomes.

We iterate the above discrete-time dynamics for a number of time steps, until the system reaches the final *static* state. As oppose to what happened with the replicator dynamics used in previous chapters, where cooperation and defection could coexist in the asymptotic state which, moreover, fluctuated in general around a well define mean value of cooperation $\langle c\rangle(b)$, now, due to this particular choice for the probability function (Eq. 6.6), the final state of the system will be one of the two absorbing states: all-C or all-D [19]. As we have seen, with this probability we allow *irrational changes of strategy*, so that a node will always have a non-zero probability of adopting the neighbor's strategy, even when the neighbor's payoff is smaller than its own. It is worth noticing that this affects the dynamics of the system in such a way that it will always end up on one of the two possible absorbing states. Therefore, one should interpret the average level of cooperation for a particular set of the parameters b/c and w, as the fraction of realizations in which the system ends up in the all-C state, (instead of the average fraction of cooperators present in the stationary state of the system).

It is worth stressing that the neighborhood that a node selects to play one round of the game with is also the one used to choose the node to compare its benefits with, but for the next round, all the nodes will select a different new effective neighborhood (except, of course, those with $k_i \leq k^*$, that play with the same opponents). This neighborhood selection procedure is quite expensive in terms of computational time. And, in addition to this, the fact that the system must achieve eventually one of the two absorbing states, makes the time evolution of the dynamics remarkably slow, specially, for the range of b/c values corresponding to intermediate values of $\langle c\rangle$.

6.2 Average Level of Cooperation

In Fig. 6.2 we plot the level of cooperation $\langle c\rangle$ as a function of the ratio b/c for different values of the restriction k^*. Obviously, as one can easily expect, the larger the value of b/c is, the cheaper being a cooperator is, and thus the larger the average level of cooperation the system can achieve. On the other hand, we have found a surprising and non-trivial dependence of the level of cooperation $\langle c\rangle$ with the value of the restriction for the number of connections k^*. From Fig. 6.2 for some low values of b/c, *i.e.*, when cooperation is relatively expensive, the largest level of cooperation is achieved when no restriction is imposed to the connectivity of the nodes, but for larger values of the ratio b/c, the opposite trend occurs, and a network with some level of connectivity restriction performs better than the original one, meaning that it achieves larger levels of cooperation. Of course, those cases with a too restrictive value for $k^* \lesssim 10$, always perform worse, regardless of the value of the ratio. Notice

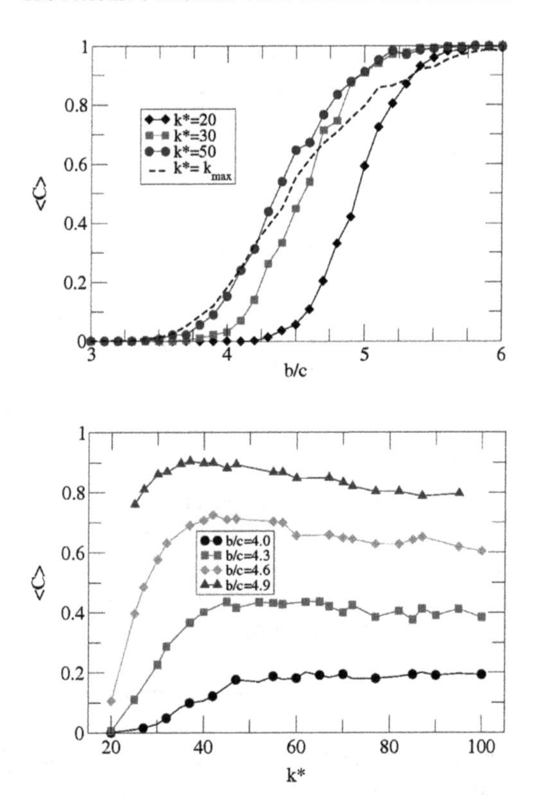

Fig. 6.2 Average level of cooperation as a function of the ratio b/c for the case of restricted number of connections *without* frustration (see Sect. 6.3 for details). The SF networks are made up of $N = 4 \times 10^3$ nodes, and the average connectivity is $\langle k \rangle = 4$. Every point is the average over 500 different realizations

Fig. 6.3 Average level of cooperation as a function of the restriction k^* for different values of the ratio b/c. The SF networks are made up of $N = 4 \times 10^3$ nodes, and the average connectivity is $\langle k \rangle = 4$. Every point is the average over 2×10^3 different realizations

that by setting $k^* = k_{max}$ we actually mean that every node i plays always with all its k_i topological neighbors.

As a matter of fact, if we represent the level of cooperation as a function of k^* for a fixed value of the ratio b/c, we obtain a non-monotonous behavior (see Fig. 6.3), where moreover, the optimum value of k^*, *i.e.* the value that yields the larger level of cooperation for a fixed b/c, seems to increase as the cooperation gets more expensive.

6.3 Imposing a More Tight Connectivity Restriction

As we have already mentioned, the first presented for the restriction of the number of interactions per node and per round of the game, k^*, is not as strict as one would like, and does not guarantee the value k^* for every node with $k_i > k^*$ present on the network. On the contrary, and due to the need of symmetry, k_i^{eff} turns out to be larger than k^*, in general. In order to obtain a more severe restriction, while preserving the symmetry condition for the interaction between nodes, we propose now a different restriction method.

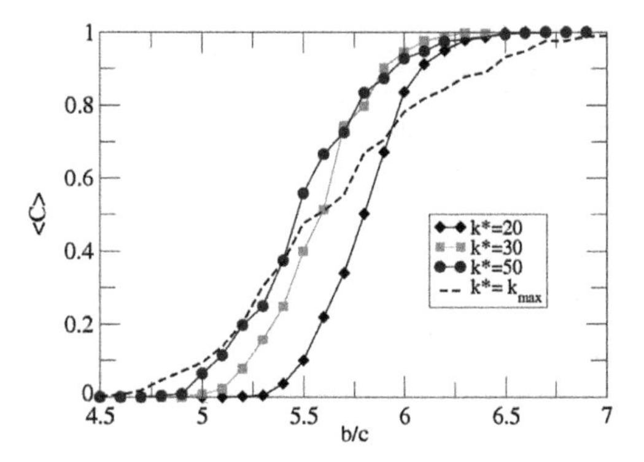

Fig. 6.4 Average level of cooperation as a function of the ratio b/c for the case of restricted number of connections *with* frustration. The SF networks are made up of $N = 4 \times 10^3$ nodes, and the average connectivity is $\langle k \rangle = 4$. Every point is the average over 500 different realizations

This second selection scheme works as follows: starting with the nodes of lower degree for a given network, we make them choose its k^* neighbors (or $k_i < k^*$ if necessary), among its topological connections, but now, we keep track of the number of possible connections still available for every node, using a tagging system, so all the nodes start with its label set to $l_i = k^*$ if $k_i > k^*$, and $l_i = k_i$ if $k_i \leq k^*$, and every time an effective connection between nodes i and j is made, we rest one unit to the labels l_i and l_j. Thus, if one node i intends to chose another node j whose label is already set to $l_j = 0$, then this pick will not be allowed, even if node i can not establish connections with anyone else. When this situation happens, we say that node i gets *frustrated*. We repeat this process for all the increasingly connected nodes, ending up with the hubs, and then, as usual, everyone plays a round of the game with its current effective neighborhood, and accumulates its benefits π_i. Then every one of them compares this value π_i with that corresponding to a neighbor, randomly chosen among its k_i^{eff}, and decides whether or not it will change its strategy with the same probability function used before. All the nodes change their strategy synchronously.

Notice that we have obviously chosen to start from the lowly connected nodes, and not the other way around in order not to margin poorly connected nodes due to the restriction procedure, so they would not get the chance to play. It is also worth mentioning that we have checked the 'average level of frustration' for the nodes on the network at a given round of the game, defined as the fraction between the sum of labels different from zero present on the system once the assignment process has finished (*i.e.* the number of connections that were not able to be established, and remain 'unused', although they were allowed), and the maximum possible number of connections the whole network would have made with the restriction k^* but without frustration. This quality always yields values under ten percent for any set of the parameters of the system. So we consider that this method, though not perfect and somehow more artificial than the first one, is a good approach to this non-trivial problem of restricting the number of connections to a constant value on a scale-free underlying topology.

Similarly to what we have done in the previous section, we show now in Fig. 6.4 the level of cooperation as a function of the ratio b/c for several values of the restriction

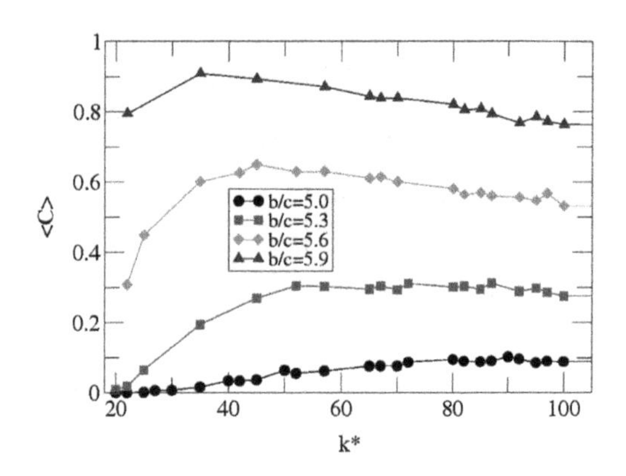

Fig. 6.5 Average level of cooperation as a function of the restriction k^* for different values of the ratio b/c for the case of restricted number of connections with frustration. The SF networks are made up of $N = 4 \times 10^3$ nodes, and the average connectivity is $\langle k \rangle = 4$. Every point is the average over 2×10^3 different realizations

k^* for the case of restricted number of connections with frustration. It can be seen that they are quite similar to those presented for the case without frustration, with mainly one quantitative difference: the value of b/c needed to maintain the same level of cooperation is larger, it is to say, the cooperation is in general more expensive in this second scenario with frustration.

But as far as the qualitative behavior is concerned, we can say that this second model behaves in the same way as the first one, so when we represent the level of cooperation as a function of k^* for a fixed value of the ratio (see Fig. 6.5), we also find a non-monotonous dependence which clearly indicates that, in order to achieve the highest level of cooperation for a fixed value of the parameters of the payoff matrix, it is better to restrict the number of interactions to a certain extent.

In order to understand better the origin of this optimum value for the number of interactions, k^*_{opt}, when playing the Prisoner's Dilemma game with costs, we will next check it for two other different scenarios: first, we will change the payoff matrix to its form without cost, and second, we will keep the cost-benefit ratio but we will adopt another updating rule, namely, the Replicator rule. By introducing these changes in our original model, we want to determine the crucial factor for the observed optimum in the number of interactions.

In this way, let us now consider the Prisoner's Dilemma game with the Fermi updating rule, but with the formulation without cost per cooperation, given by the following payoff matrix:

$$\begin{array}{c} \\ C \\ D \end{array} \begin{array}{c} \begin{array}{cc} C & D \end{array} \\ \begin{pmatrix} R & S \\ T & P \end{pmatrix} \end{array} = \begin{array}{c} \\ C \\ D \end{array} \begin{array}{c} \begin{array}{cc} C & D \end{array} \\ \begin{pmatrix} 1 & 0 \\ b & 0 \end{pmatrix} \end{array} \tag{6.7}$$

where we fix, as usual, $R = 1$ and $P = S = 0$. In Fig. 6.6 we show the average level of cooperation in the system as a function of the restriction k^*, for different values of the temptation to defect, b. In this case, we can clearly see that, for any fixed value of b, the system renders the highest value of cooperation for the unrestricted situation *i.e.*, for $k^* = 4 \times 10^3$ (not explicitly shown). So, comparing Fig. 6.6 with Figs. 6.5 or

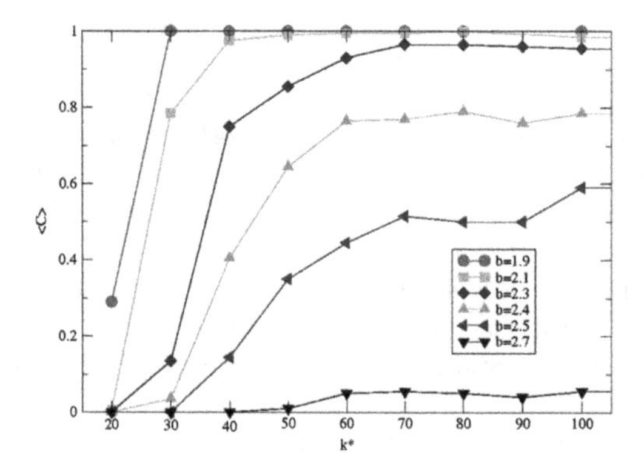

Fig. 6.6 Average level of cooperation as a function of the matching limitation, k^*, for the case of restriction in the number of connections without frustration, and using the Fermi updating rule and the formulation of the Prisoner's Dilemma without cost for cooperation. The SF networks are made up of $N = 4 \times 10^3$ nodes, and the average connectivity is $\langle k \rangle = 4$. Every point is the average over 200 different realizations

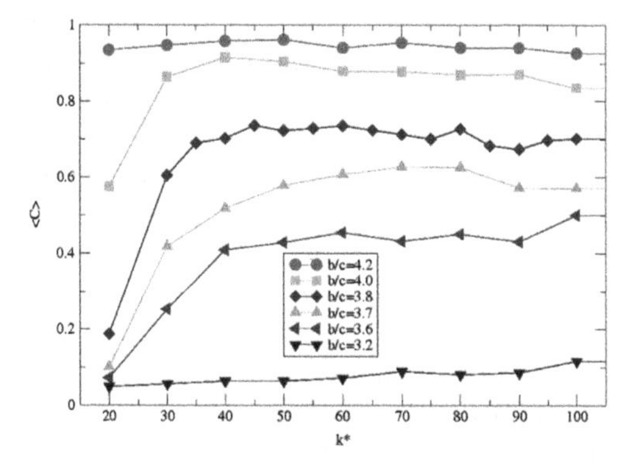

Fig. 6.7 Average level of cooperation as a function of the mate limitation, k^*, for the case of a restriction in the number of connections without frustration, and using the Replicator updating rule and the formulation of the Prisoner's Dilemma with cost for cooperation. The SF networks are made up of $N = 4 \times 10^3$ nodes, and the average connectivity is $\langle k \rangle = 4$. Every point is the average over at least 200 different realizations

6.3, we can conclude that the reason why such an optimum, k^*_{opt}, exists is due to a necessary compromise every node has to establish between the cost of cooperating with all its neighbors and the benefits obtained in those interactions. It is reasonable to think that, even if all neighbors are cooperators, it will be very expensive to pay the cost for cooperating with all of them, so the benefits will decrease. On the other hand, if one interacts with too few of its neighbors, the cost will be low, but so will be the benefit.

Finally, as a further check, let us consider the second change to our model: the Prisoner's Dilemma game with cost and the Replicator updating rule. We show in Fig. 6.7 the result of our simulations, and we can see that the optimum, k^*_{opt}, reappears in this scenario, though it is not so pronounced as in the case with Fermi-like updating rule for any value of the ratio b/c. We can conclude that the root of this optimum is indeed in the use of a cost formulation of the Prisoner's Dilemma.

6.4 Conclusions

In this chapter we have studied a realistic -but almost unexplored until now- scenario where the number of interactions that a node can establish per round of the game are restricted to a maximum value k^*, regardless of its topological connectivity of the nodes. We have studied two different mechanisms to perform such restriction. The first method does not need any global information, since every node chooses its k^* game mates and it just guarantees the symmetry of the interactions. However, as it turned out, this is not a very strict restriction, since the actual connectivity of some of the nodes is in general $k_i^{eff} \gtrsim k^*$. The second one is somehow more artificial, since one needs some global knowledge of the network (precisely the degree of every node) in order to proceed. But on the other hand, it strictly imposes the restriction of having k^* game mates.

We have studied the effect of such restrictions on scale-free networks and found that the results are qualitatively the same for both methods. In particular, we have focus on the level of cooperation achieved by the system at the stationary state, comparing the results with those for the Prisoner's Dilemma game on the original BA scale-free networks. Our main result is that for a range of values of the cost-benefit b/c ratio of the payoff matrix, the highest levels of cooperation are achieved when some connectivity restriction is imposed on the network, *i.e.*, the larger levels of cooperation do not occur for the original unrestricted BA scale-free network scenario, but for a more realistic situation, where every node can engage on a round of the game just with a certain number of neighbors k^* that is, in general, lower than its actual topological connectivity, k_i.

This is a quite surprising result, since previous studies always have pointed out the well-known enhancement of cooperation due to heterogeneity of the underlying topology. Here we have clarified that this is only true up to a certain extent: although heterogeneity does greatly favor cooperation when comparing it with the case of random networks, the restriction of forcing the nodes to play just with $k^* < k_i$ of its neighbors in every round of the game seems to lead to even larger levels of cooperation in some regions of the parameter space of the ratio b/c.

We also showed that the existence of this optimum, k^*_{opt}, was due to the compromise between the cost of cooperating with all its neighbors and the benefits obtained from those interactions. In order to confirm this hypothesis, we simulated the dynamics for two other scenarios: in the first one, we kept the updating rule, but we changed the formulation of the dilemma, using a payoff matrix where the cost per cooperation is zero. As we expected, now the highest values of cooperation achieved by the system occur when there is no limitation to the number of interactions. On the other hand, if we consider the dynamics with a different updating rule, namely the Replicator rule, but we keep the cost-benefit ratio formulation, then the optimum value k^*_{opt} appears again. In conclusion, the results shown in this chapter point out that the particular formulation chosen when implementing the Prisoner's Dilemma on top of complex topologies will introduce important differences in the outcome of the dynamics, specially in realistic scenarios as the one proposed here.

References

1. M. A. Nowak and R. M. May, Nature **359**, 826 (1992).
2. E. Lieberman, C. Hauert, and M. A. Nowak, Nature **433**, 312 (2005).
3. H. Ohtsuki, E. L. C. Hauert, and M. A. Nowak, Nature **441**, 502 (2006).
4. V. M. Eguíluz, M. G. Zimmermann, C. J. Cela-Conde, and M. San Miguel, American Journal of Sociology **110**, 977 (2005).
5. F. C. Santos and J. M. Pacheco, J. Evol. Biol. **19**, 726 (2006).
6. M. Newman, SIAM Review **45**, 167 (2003).
7. S. Boccaletti, V. Latora, Y. Moreno, M. Chavez, and D. Hwang, Phys. Rep. **424**, 175 (2006).
8. F. C. Santos and J. M. Pacheco, Phys. Rev. Lett. **95**, 098104 (2005).
9. F. C. Santos, J. M. Pacheco, and T. Lenaerts, Proc. Natl. Acad. Sci. USA **103**, 3490 (2006).
10. F. C. Santos, F. J. Rodrigues, and J. M. Pacheco, Proc. Biol. Sci. **273**, 51 (2006).
11. F. Santos, M. Santos, and J. Pacheco, Nature **454**, 231 (2008).
12. A. Barabási and R. Albert, Science **286**, 509 (1999).
13. J. M. Pacheco, A. Traulsen, and M. A. Nowak, Phys. Rev. Lett. **97**, 258103 (2006).
14. H. Ohtsuki and M. A. Nowak, J. Theor. Biol. **243**, 86 (2006).
15. L. E. Blume, Games and Economic Behavior **5**, 387 (1993).
16. G. Szabó and C. TOke, Phys. Rev. E **58**, 69 (1998).
17. C. Hauert and G. Szabó, Am. J. Phys. **73**, 405 (2005).
18. A. Traulsen, J. M. Pacheco, and M. A. Nowak, J. Theor. Biol. **246**, 522 (2007).
19. A. Traulsen, M. Nowak, and J. Pacheco, Phys. Rev. E **74**, 011909 (2006).

Part II
Evolutionary Dynamics on Growing Complex Networks

Presentation of Part II

In this second part of the Thesis, we will focus on the study of the coupling between the growth of a complex topology and the dynamics taking place simultaneously on top of it.

As we have been seeing, a great deal of effort has been aimed to study the influence of a (static) complex topologies on the outcome of several games [1–14]. Specially the PD, being a paradigmatic example of cooperative-defective interaction, has been proved to be a very useful tool when trying to explain the reasons why such a expensive behavior as cooperation can arise and survive in a population. On the other hand, it has been proved for many real networked systems in a wide variety of contexts that topology greatly affects dynamics but also the other way around ([15] and references therein), establishing thus a feedback loop. In this way, when it comes specifically to Evolutionary Game Theory on non-static graphs, some nice works [16–19] have tried to consider a more complex situation, as far as the structure is concerned, by placing the dynamics on a N-sized network whose links are being rewired, according to some dynamics-dependent rules (adaptative networks), or even using two different networks, one for the interaction, the other one for the comparison procedure. Nonetheless, to our knowledge, the attempt we have made is the first to aim a growing structure, where this growth is entangled somehow with the dynamics of the nodes. We have developed two models to address this issue, and in both of them the particular dynamics evolving in the population is the PD game. However, there are some important differences between the specifics of each one.

Thus, in Chap. 7 we introduce the first model, for which we will consider that the probability of attachment is a linear function of the fitness of the chosen node. On the other hand, the strategy updating rule we will use is Replicator-like. During this chapter, we will study the different topologies that can arise depending on the values of the relevant parameters of the system. Specifically, we will be able to build random and SF networks. We will study the dynamical organization of cooperation among connectivity classes for heterogeneous structures obtained with our model, comparing these results with the well-known ones for SF BA networks, and trying to explain the differences found. Also, we will check the average level of cooperation achieved by our networks, in two instants: when the growth has just stopped, and some time later, after letting the population play the same game, but without adding new individuals. We will show that the structures built via this first model can support, when used as static substrate for the PD game, higher levels of cooperation than the celebrated BA SF networks [3–5]. Besides, we will compare these levels of cooperation with those for a rewired version of the resultant topology, and we will be able to make some conclusions about the adequacy of the networks our model gives rise to, when it comes to supporting cooperation. Moreover, we have found that the structures obtained with this model share some topological features with real systems, such as the power-law dependence of the clustering coefficient with the degree of the nodes, compatible with hierarchical organizations. So we consider that our work can help understand the origin of these heterogeneous networks from an evolutionary point of view.

In Chap. 8 we propose a second model, that is somehow different from the first one, but always within the framework of an interdependence between the growth and the dynamics. Thus, we consider again that the nodes are playing the PD game, although with another formulation in terms of the payoff matrix. Also, the strategy updating rule is dictated by a Fermi-like function, which allows irrational changes of strategy, it is to say, it is possible to imitate a neighbor with a worse payoff. As we will see, the introduction of this Fermi probability will affect greatly the final state of the system, when it comes to the levels of cooperation. Moreover, the probability of attachment we will use in this second model is exponential with the fitness of the nodes, instead of linear, which permits the appearing of not only random and scale-free structures, buy also star-like ones, with a few nodes that are 'super-hubs'. Apart from the degree distribution and the final levels of cooperation in the system, we are also interested in analyzing whether cooperation benefits from the growth process or just from the resulting complex structure, and to that aim, we will look again into both the level of cooperation after finishing the growth and after letting the system evolve for some time. We will also consider the case of using the full grown network as a static substrate, and letting the dynamics evolve after reinitializing the level of cooperation to 50 % of each strategy, randomly distributed. In this department, we will find some remarkable differences between the two models, since for this second one cooperation turns out not to get promoted when using the resulting topologies as static substrate for the dynamics.

References

1. G. Szabó and G. Fáth, Phys. Rep. **446**, 97 (2007)
2. M. Nowak, Science **314**, 1560 (2006)
3. F. C. Santos and J. M. Pacheco, Phys. Rev. Lett. **95**, 098104 (2005)
4. F. C. Santos, F. J. Rodrigues, and J. M. Pacheco, Proc. Biol. Sci. **273**, 51 (2006)

5. F. C. Santos and J. M. Pacheco, J. Evol. Biol. **19**, 726 (2006)
6. F. C. Santos, J. M. Pacheco, and T. Lenaerts, Proc. Natl. Acad. Sci. USA **103**, 3490 (2006)
7. H. Ohtsuki, E. L. C. Hauert, and M. A. Nowak, Nature **441**, 502 (2006)
8. G. Abramson and M. Kuperman, Phys. Rev. E **63**, 030901(R) (2001)
9. V. M. Eguíluz, M. G. Zimmermann, C. J. Cela-Conde, and M. San Miguel, American Journal of Sociology **110**, 977 (2005)
10. T. Killingback and M. Doebeli, Proc. R. Soc. Lond. **263**, 1135 (1996)
11. A. Szolnoki, M. Perc, and Z. Danku, Physica A **387**, 2075 (2008)
12. J. Vukov and G. S. A. Szolnoki, Phys. Rev. E **77**, 026109 (2008)
13. J. Gómez-Gardeñes, M. Campillo, L. M. Floría, and Y. Moreno, Phys. Rev. Lett. **98**, 108103 (2007)
14. R. Jiménez, H. Lugo, J. Cuesta, and A. Sánchez, J. Theor. Biol. **250**, 475 (2008)
15. T. Gross and B. Blasius., J. R. Soc. Interface **5**, 259 (2008)
16. F. C. Santos, J. M. Pacheco, and T. Lenaerts, PLos Comput. Biol. **2**(**10**), e140 (2006)
17. J. M. Pacheco, A. Traulsen, and M. A. Nowak, Phys. Rev. Lett. **97**, 258103 (2006)
18. M. G. Zimmermann, V. M. Eguiluz, and M. S. Miguel, Phys. Rev. E **69**, 065102(R) (2004).
19. H. Ohtsuki, M. A. Nowak, and J. M. Pacheco, Phys. Rev. Lett. **98**, 108106 (2007)

Chapter 7
Complex Networks from Evolutionary Preferential Attachment

In this chapter we analyze the growth and formation of complex networks by *coupling* the network formation rules to the dynamical states of the elements of the system. As we have already mentioned, some mechanisms have been proposed for constructing complex scale-free networks similar to those observed in natural, social and technological systems from purely topological arguments (for instance, using a preferential attachment rule or any other rule available in the literature [1, 2]). As those works do not include information on the specific function or origin of the network, it is very difficult to discuss the origin of the observed networks on the basis of those models, hence motivating the question we are going to address. The fact that the existing approaches consider separately the two directions of the feedback loop between the function and form of a complex system demands for a new mechanism where the network grows coupled to the dynamical features of its components. Our aim here is to introduce for the first time an attempt in this direction, by linking the growth of the network to the dynamics taking place among its nodes.

Our model combines two ideas in a novel manner: preferential attachment and evolutionary game dynamics. Indeed, with the problem of the emergence of cooperation as a specific application in mind, we consider that the nodes of the network are individuals involved in a social dilemma and that newcomers are preferentially linked to nodes with high fitness, the latter being proportional to the payoffs obtained in the game. In this way, the fitness of an element is not imposed as an external constraint [3, 4], but rather it is the result of the dynamical evolution of the system. At the same time, the network is not exogenously imposed as a static and rigid structure on top of which the dynamics evolves, but instead it grows from a small seed and acquires its structure during its formation process. Finally, we stress that this is not yet another preferential attachment model, since the quantity that favors linking to the new nodes has no direct relation with the instantaneous topology of the network. In fact, as we will see, the main result of this interplay is the formation of homogeneous or heterogeneous networks (depending on the values of the parameters of our system) that share a number of topological features with real world networks such as a high clustering and degree–degree correlations. Thus, the model we propose not

J. Poncela Casasnovas, *Evolutionary Games in Complex Topologies*, Springer Theses, 117
DOI: 10.1007/978-3-642-30117-9_7, © Springer-Verlag Berlin Heidelberg 2012

only explains why heterogeneous networks are appropriate to sustain cooperation, but also provides an evolutionary mechanism for their origin. On the other hand, we will find that there are some important and quite surprising differences between the networks we build using this model, and SF topologies, as far as the microscopic organization of the dynamics is concerned.

7.1 The Model

Our model naturally incorporates an intrinsic feedback between dynamics and topology. In this way, the growth of the network starts at time $t = 0$ with a core of m_0 fully connected nodes, whose initial strategy is cooperation. New elements are incorporated to the network and attached to m existing nodes with a probability that depends on the payoff of each node. On the other hand, the particular dynamics we consider is dictated by the Prisoner's Dilemma (PD) game [5]. Initially, every node adopts with the same probability one of the two available strategies, cooperation C or defection D. At equally spaced time intervals (denoted by τ_D) each node i of the network plays with its $k_i(t)$ neighbors and the obtained payoffs are considered to be the measure of its evolutionary fitness, $f_i(t)$. There are three possible situations for each pair of nodes linked together in the network, as far as the outcome of the game is concerned: *(i)* if two cooperators meet, both receive R, when *(ii)* two defectors play, both receive P, while *(iii)* if a cooperator and a defector compete, the former receives S and the latter obtains T. The ordering of the four payoffs is the following: $T = b > R = 1 > P = S = 0$, where we haver fixed the value of the three parameters as usual [6–8], when considering the weak Prisoner's Dilemma game (see Chap. 3). Thus, the temptation to defect b remains as the unique free parameter of the dynamics. After playing, every node i compares its evolutionary fitness (payoff) with that corresponding to a randomly chosen neighbor j. Then, if $f_i(t) \geq f_j(t)$, node i keeps its strategy for the next round of the game, but if $f_j(t) > f_i(t)$ node i adopts the strategy of player j with probability [8–14]

$$P_i = \frac{f_j(t) - f_i(t)}{b \cdot \max\left[k_i(t), k_j(t)\right]} . \tag{7.1}$$

The growth of the network proceeds by adding a new node with m links to the preexisting ones at equally spaced time intervals (denoted by τ_T), and the probability that a node i in the network receives one of the m new links is

$$\Pi_i(t) = \frac{1 - \epsilon + \epsilon f_i(t)}{\sum_{j=1}^{N(t)}(1 - \epsilon + \epsilon f_j(t))} , \tag{7.2}$$

where $N(t)$ is the size of the network at time t, and the parameter $\epsilon \in [0, 1)$ controls the weight of the fitness $f_i(t)$ [15] during the growth of the network. Provided that $\epsilon > 0$, nodes with $f_i(t) \neq 0$ are preferentially chosen.

The growth of the network as defined above is thus linked to the evolutionary dynamics that is simultaneously evolving in the system, and it is controlled on the one hand by the parameter ϵ, but also by the two time scales, τ_T and τ_D, associated to both processes. Therefore, Eq. 7.2 can be viewed as an '*Evolutionary Preferential Attachment*' (EPA) mechanism. Depending on the value of ϵ, we can have two extreme situations:

(i) When $\epsilon \simeq 0$, referred to as the *weak selection limit* [16], the network growth is almost independent of the evolutionary dynamics as all nodes have roughly the same probability of attracting new links.
(ii) Alternatively, in the *strong selection limit*, $\epsilon \to 1$, the fittest players (highest payoffs) are much more likely to attract the links from newcomers.

Between the above situations, there is a continuum of intermediate values that will give rise to a wide range of in-between behaviors.

We have carried out numerical simulations of the model exploring the (ϵ, b) space. It is worth mentioning that we have also explored different time relations τ_D/τ_T, but for the time being, we focus on the results obtained when $\tau_D/\tau_T > 1$, namely, the network growth is faster than the evolutionary dynamics. Later on we will discuss the effects associated to other time ratios. Taking $\tau_T = 1$ as the reference time, networks are generated by adding nodes every time step, while they play at discrete times given by τ_D. As $\tau_D > \tau_T$, the linking procedure is done with the payoffs obtained the last time the nodes played. All results reported have been averaged over at least 10^3 realizations, and the number of links of a newcomer is taken to be $m = 2$ (so the average connectivity will be $\langle k \rangle = 2m = 4$), whereas the size of the initial core is $m_0 = 3$.

7.2 Degree Distribution and Average Level of Cooperation

The dependence of the degree distribution on ϵ and b is shown in Fig. 7.1. As it can be seen, the weak selection limit produces homogeneous networks characterized by a tail that decays exponentially with k. Alternatively, when ϵ is large, scale-free networks arise. Although this might a priori be expected from the definition of the growth rules, this needs not be the case: indeed, it must be taken into account that in a one-shot PD game, defection is the best strategy regardless of the opponent's strategy. However, if the network dynamics evolves into a state in which all players (or a large part of the network) are defectors, they will often play against themselves and their payoffs will be reduced (we recall that $P = 0$). The system's dynamics will then end up in a state close to an all-D configuration, rendering $f_i(t) = 0 \; \forall i \in [1, N(t)]$ in Eq. 7.2. From this point on, new nodes would attach randomly to other existing nodes (see Eq. 7.2) and therefore no hubs can come out. This turns out not to be the case, which indicates that for having some degree of heterogeneity, a nonzero level of cooperation is needed. Conversely, the heterogeneous character

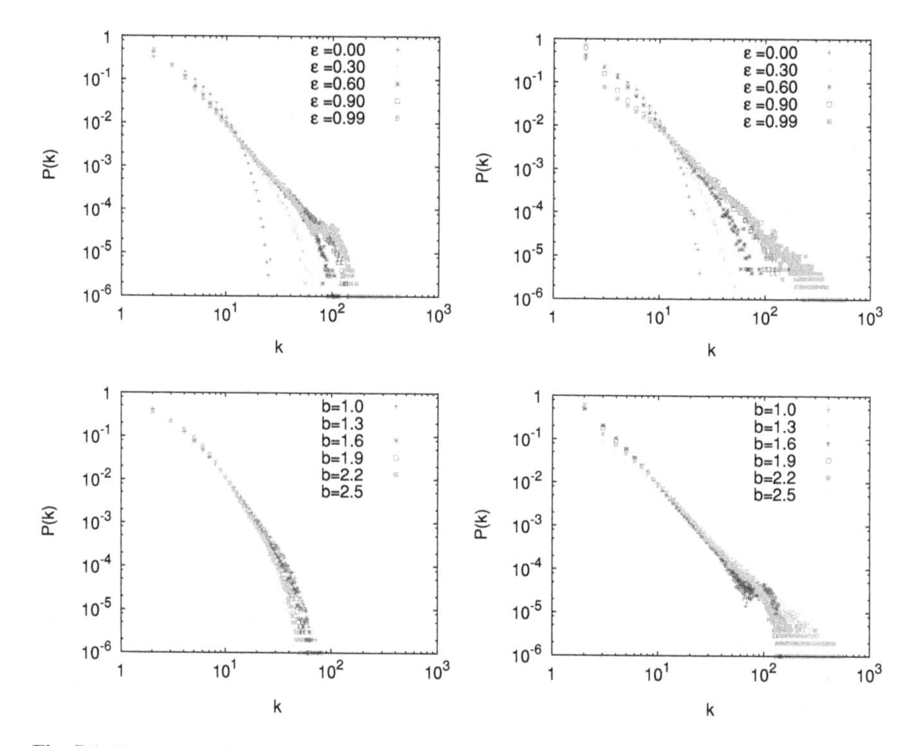

Fig. 7.1 Degree distribution of the topologies created for fixed values of $b = 1.5$ (*Top left*) and $b = 2.5$ (*Top right*), and fixed values of $\epsilon = 0.3$ (*Bottom left*) and $\epsilon = 0.99$ (*Bottom right*). The networks are made up of $N = 10^3$ nodes, with average connectivity $\langle k \rangle = 4$, and $\tau_D = 10\tau_T$. Every point is the average of 300 independent realizations

of the system provides a feedback mechanism for the survival of cooperators that would not overcome defectors otherwise.

In Fig. 7.1 we also show the dependence of the degree of heterogeneity of the networks with the temptation to defect, and we found out that only in the strong selection limit, it depends slightly on b. On the other hand, for small values of ϵ, there is not any dependence of the degree distribution on b, because in this scenario the dynamics does not play a relevant role on the attachment, on the contrary, it is almost random.

Regarding the outcome of the dynamics, we have also represented the average level of cooperation $\langle c \rangle$, as a function of the two model parameters ϵ and b. The Fig. 7.2 shows that as ϵ grows for a fixed value of $b \gtrsim 1$, the level of cooperation increases. In particular, in the strong selection limit $\langle c \rangle$, the system attains its maximum value. This is a somewhat counterintuitive result as in the limit $\epsilon \to 1$, new nodes are preferentially linked to those with the highest payoffs, which for the PD game, should correspond to defectors. However, the population achieves the highest value of $\langle c \rangle$. On the other hand, higher levels of cooperation are achieved in heterogeneous rather than in homogeneous topologies, which is consistent with previous findings [8, 14, 17].

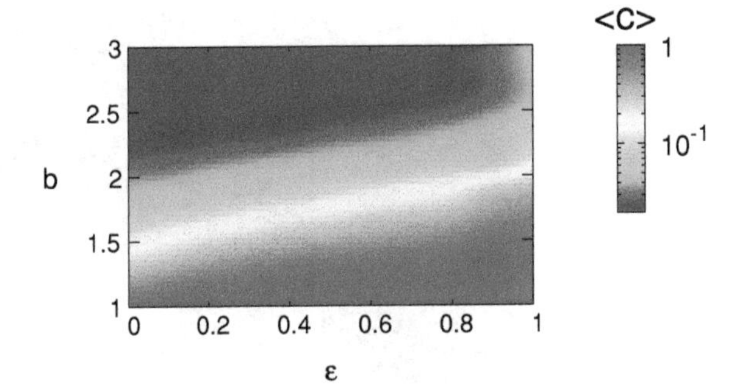

Fig. 7.2 Color-coded average level of cooperation in the system $\langle c \rangle$ right at the end of the EPA procedure, it is to say, when the final size is achieved as a function of the temptation to defect b and the selection pressure ϵ. The networks are made up of 10^3 nodes with average connectivity $\langle k \rangle = 4$ and $\tau_D = 10\tau_T$

7.3 Degree Distribution Among Cooperators

In this section we want to study the dependence between strategy and degree of connectivity, comparing this results with those obtained for the static SF scenario, where we recall that cooperators occupy always the highest and medium classes of connectivity, while defectors are not stable when setting on the hubs (Sect. 3.5). As we will show, the interplay between the local structure of the network and the hierarchical organization of cooperation seems to be highly nontrivial, and quite different from what has been reported for static scale-free networks [8, 14]. In Fig. 7.3 one can see that, surprisingly enough, as the temptation to defect increases, the likelihood that cooperators occupy the hubs decreases. Indeed, during network growth, cooperators are not localized neither at the hubs nor at the lowly connected nodes, but in intermediate degree classes. It is important to realize that this is a new effect that arises from the competition between network growth and the evolutionary dynamics. In particular, it highlights the differences between the microscopic organization in the steady state for the PD game in static networks and that found when the network is evolving.

To address this interesting and previously unobserved phenomenon, we have developed a simple analytical argument that can help understand the reasons behind it. Let k_i^c be the number of cooperator neighbors of a given node i. Its fitness is $f_i^d = bk_i^c$ if node i is a defector, and $f_i^c = k_i^c$ if it is a cooperator. The value of k_i^c is expected to change because of two factors. On the one hand, due to the network growth (node accretion flow, at a rate of one new node each time unit τ_T) and on the other hand, due to imitation processes dictated by Eq. 7.1, that take place at a pace τ_D. As it has been mentioned before, we will focus on the case in which τ_D is much larger than τ_T, for now. Thus, the expected increase of fitness is:

$$\Delta f_i = \Delta_{flow} f_i + \Delta_{evol} f_i, \tag{7.3}$$

where $\Delta_{flow} f_i$ stands for the variation of fitness in node i due to the newcomers flow, and $\Delta_{evol} f_i$ is the change in fitness due to changes of neighbors' strategies. The above expression leads to an expected increase in k_i^c given by:

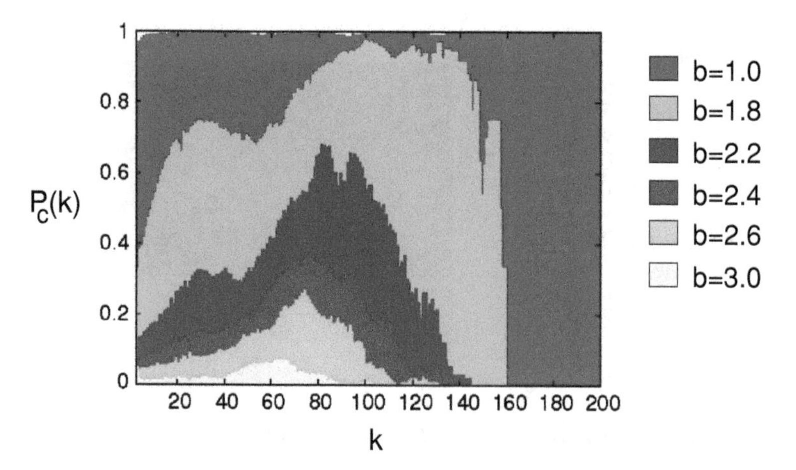

Fig. 7.3 Probability $P_c(k)$ that a node with connectivity k plays as a cooperator for different values of b in the strong selection limit ($\epsilon = 0.99$) at the end of the growth of a network with $N = 10^3$ nodes and average connectivity $\langle k \rangle = 4$

$$\Delta k_i^c = k_i^c(t + \tau_D) - k_i^c(t) = \Delta_{flow} k_i^c + \Delta_{evol} k_i^c. \tag{7.4}$$

On the other hand, the expected increase of degree of node i in the interval of time $(t, t + \tau_D)$ only has the contribution from newcomer flow, and recalling that new nodes are generated with the same probability to be cooperators or defectors, i.e, $\rho_0 = 0.5$, it will take the form:

$$\Delta k_i = \Delta_{flow} k_i = 2\Delta_{flow} k_i^c. \tag{7.5}$$

If the fitness (hence connectivity) of node i is high enough to attract a significant part of the newcomer flow, the first term in Eq. 7.3 dominates at short time scales, and then the hub's degree k_i increases exponentially. Connectivity patterns are then dominated by the growth by preferential attachment, ensuring, as in the BA model [18], that the network will have a SF degree distribution. Moreover, the rate of increase of the connectivity:

$$\Delta_{flow} k_i^c = \frac{1}{2} m \tau_D \frac{f_i}{\sum_j f_j} \tag{7.6}$$

is larger for a defector hub by a factor b, because of its larger fitness, and then one should expect hubs to be mostly defectors, as confirmed by the results shown in Fig. 7.3. This small set of most connected defector nodes attracts most of the newcomer flow.

On the contrary, for nodes of intermediate degree, say of connectivity $m \ll k_i \ll k_{max}$, the term $\Delta_{flow} f_i$ in Eq. 7.3 can be neglected, in other words, the arrival of new nodes is a rare event, so for a large time scale, we have that $\dot{k}_i = 0$. Note that

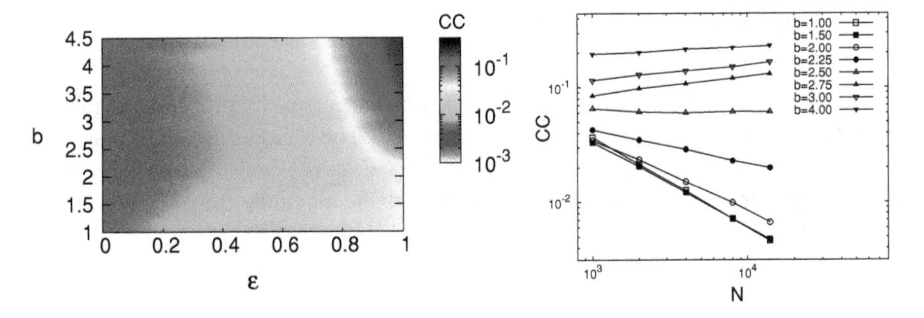

Fig. 7.4 *(Left)* Clustering coefficient CC as a function of b and ϵ. *(Right)* Scaling of CC with the network size for several values of b in the strong selection limit ($\epsilon = 0.99$). The networks are made up of $N = 10^3$ nodes and have average connectivity $\langle k \rangle = 4$

if $\dot{k}_i(t) = 0$ for all t in an interval $t_0 \leq t \leq t_0 + T$, the size of the neighborhood is constant during that whole interval T, and thus the evolutionary dynamics of strategies through imitation is exclusively responsible for the strategic field configuration in the neighborhood of node i. During these periods, the probability distribution of strategies in the neighborhood of node i approaches that of a static network. Thus, recalling that the probability for this node i of intermediate degree to be a cooperator is large in the static regime [14] (see also Sect. 3.5), we then arrive to the conclusion that for these nodes the density of cooperators must reach a maximum, in agreement with Fig. 7.3. Of course, it is clear that this scenario can be occasionally subject to sudden avalanche-type perturbations following "punctuated equilibrium" patterns in the rare occasions in which a new node arrives.

Furthermore, our simulations show that these features of the shape of the curve $P_c(k)$ are indeed preserved as time goes by, giving further support to the above argument based on time scale separation and confirming that our understanding of the mechanisms at work in the model is correct.

7.4 Clustering Coefficient and Degree–Degree Correlations

Apart from the degree distribution, we are also interested in exploring other topological features emerging from the interaction between network growth and the evolutionary dynamics in our EPA networks. Namely, we will focus on two important topological measures that describe the existence of nontrivial two-body an three-body correlations: the degree–degree correlations and the clustering coefficient respectively. We will show that the networks generated by the EPA model display both hierarchical clustering and disassortative degree–degree correlations.

7.4.1 Clustering Coefficient

The clustering coefficient of a given node i, cc_i, expresses the probability that two neighbors j and m of node i, are also connected. The value of cc_i is obtained by counting the actual number of edges, denoted by e_i, in $\mathcal{G}_i$, the subgraph induced by the k_i neighbors of i, and dividing this number by the maximum possible number of edges in $\mathcal{G}_i$:

$$cc_i = \frac{2e_i}{k_i(k_i - 1)} . \tag{7.7}$$

The clustering coefficient of a given network, CC is calculated by averaging the individual values $\{cc_i\}$ across the network nodes, $CC = \sum_i cc_i/N$. Therefore, the clustering coefficient CC measures the probability that two different neighbors of a same node, are also connected to each other. In the left panel of Fig. 7.4 we show the value of CC as a function of b and ϵ. In this figure we observe that it is in the strong selection limit where the largest values of CC are obtained. Therefore, in this regime, not only highly heterogeneous networks are obtained but the nodes also display a large clusterization into neighborhoods of densely connected nodes. In the right panel of Fig. 7.4 we show the scaling of the clustering with the network size $CC(N)$ in the strong selection limit. In this case we observe that for $b \geq 2.5$ the value of CC is stationary while when $b < 2.5$ the addition of new nodes in the network tends to decrease its clustering.

We now focus on the dependence of the clustering coefficient CC with the degree of the nodes, k, in the strong selection limit ($\epsilon = 0.99$). Interestingly enough, we show in Fig. 7.5 that the dependence of $CC(k)$ is consistent with a hierarchical organization, and it can be approximately expressed by the power law $CC(k) \sim k^{-\beta}$, a statistical feature found to describe many real-world networks [2]. The behavior of $CC(k)$ in Fig. 7.5 can be understood by recalling that in scale-free networks, cooperators are not extinguished even for large values of b if they organize into clusters of cooperators that provide the group with a stable source of benefits [14]. But to understand this feature in detail, let us assume that a new node j arrives to the network: since the attachment probability depends on the payoff of the receiver, node j may link either to a defector hub or to a node belonging to a cooperator cluster. In the first scenario, it will receive less payoff than the hub, so it will sooner or later imitate its strategy, and therefore will get trapped playing as a defector with a payoff equal to $f_j = 0$. Thus, node j will not be able to attract any links during the subsequent network growth. On the other hand if it attaches to a cooperator cluster and forms a triad with m elements of the cooperator cluster, two different outcomes are possible, depending on its initial strategy: if it plays as a defector, the triad may eventually be invaded by defectors an may end up in a state where the nodes have no capacity to receive new links. But if it plays as a cooperator, the group will be reinforced, both in its robustness against defector attacks and in its overall fitness to attract new links.

To sum up, playing as a cooperator while taking part in a successful (high fitness) cooperator cluster reinforces its future success, while playing as a defector under-

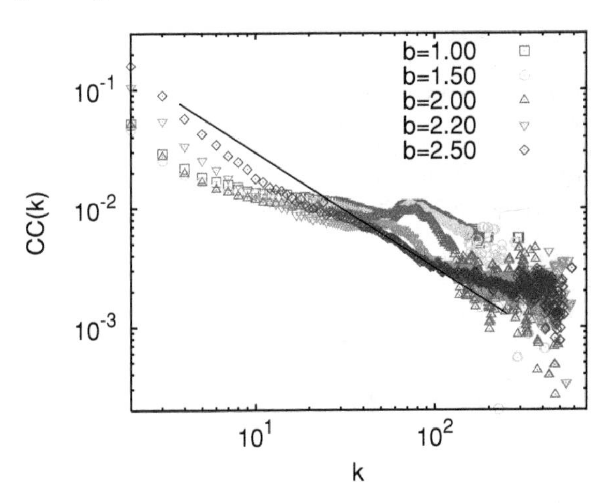

Fig. 7.5 Dependence of the clustering coefficient $CC(k) \sim k^{-\beta}$ with the nodes' degree for different values of b in the strong selection limit ($\epsilon = 0.99$). The networks are made up of $N = 10^3$ nodes and average connectivity $\langle k \rangle = 4$. The straight line is an eye guide that corresponds to k^{-1}

mines its future fitness and leads to dynamically and topologically frozen structures (it is to say, with $f_i = 0$), so defection cannot take long-term advantage from cooperator clusters. Therefore, cooperator clusters that emerge from cooperator triads to which new cooperators are attached can then continue to grow if more cooperators are attracted or even if defectors attach to the nodes whose connectivity verifies $k > mb$. Moreover, the stability of cooperator clusters and its global fitness grow with their size, specially for their members with higher degree, and naturally favors the formation of triads among its components. Thus, it follows from the above mechanism that a node of degree k is a vertex of $(k-1)$ triangles, and then

$$CC(k) = \frac{k-1}{k(k-1)/2} = 2/k\ , \tag{7.8}$$

which is exactly the sort of functional form for the clustering coefficient we have found (Fig. 7.5).

7.4.2 Degree–Degree Correlations

Now we turn our attention to the degree–degree correlations of EPA networks. Degree–degree correlations are defined by the conditional probability, $P(k^{'}|k)$, that a node of degree k is connected with a node of degree $k^{'}$. However, since the computation of this probability yields very noisy results, it is difficult to assess whether degree–degree correlations exist in a given network topology. A useful measure to overcome this technical difficulty is to compute the average degree of the neighbors of nodes with degree k, $K_{nn}(k)$, that is related with the probability $P(k|k^{'})$ as

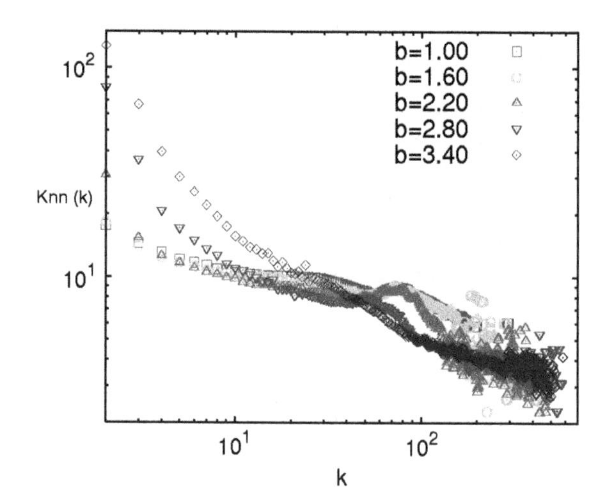

Fig. 7.6 Degree–degree correlations among the nodes of the EPA networks. We plot the average nearest-neighbors degree $K_{nn}(k)$ of a node of degree k for several values of the parameter b used to generate the networks. The networks are generated with $\epsilon = 0.99$, and have $N = 4 \cdot 10^3$ nodes and average connectivity $\langle k \rangle = 4$. Note that negative correlations imply that hubs are not likely to be connected to each other

$$K_{nn}(k) = \sum_{k'} k' P(k'|k) \,. \tag{7.9}$$

In networks without degree–degree correlations the function $K_{nn}(k)$ is flat whereas for degree–degree correlated networks the function is approximated by $K_{nn} \sim k^{\nu}$ and the sign of the exponent ν reveals the nature of the correlations. For assortative networks $\nu > 0$ and nodes are connected to neighbors with similar degrees. On the other hand, for disassortative networks, $\nu < 0$, and high degree nodes tend to be surrounded by low degree nodes.

In Fig. 7.6 we plot several functions $K_{nn}(k)$ corresponding to different values of b in the strong selection limit. We observe that for all the cases exists a negative correlation that makes highly connected nodes to be more likely connected to poorly connected nodes and vice versa. Therefore the EPA topologies are disassortative, and this behavior is enhanced as the temptation to defect, b, increases as observed from the slope of the curves in the log-log plot. This disassortative nature of EPA networks will be of relevance when analyzing the results presented in the following section.

7.5 Dynamics on Static Networks Constructed Using the EPA Model

Up to this section we have analyzed the topology and the dynamics of the EPA networks while the growing process is still going on. Now we adopt a different perspective by considering the networks as static substrates while studying the evolutionary dynamics of the nodes. This approach will be done in different ways allowing us to have a deeper insight on the EPA networks and their robustness.

7.5.1 Stopping Growth and Letting the Evolutionary Dynamics Evolve

To confirm the robustness of the networks generated by Evolutionary Preferential Attachment, let us consider the realistic situation that after incorporating a large number of participants, the network growth stops when a given size N is reached, and after that, only evolutionary dynamics takes place. The question we aim to address here is: will the cooperation observed during the co-evolution process resist?

In Fig. 7.7, we compare the average level of cooperation $\langle c \rangle$ when the network just ceased growing with the same quantity computed after allowing the evolutionary dynamics to evolve many more time steps without attaching new nodes, $\langle c \rangle_\infty$. The green area indicates the region of the parameter b where the level of cooperation increases with respect to that at the moment the network stops growing. On the contrary, the red zone shows that beyond a certain value, b_c, of the temptation to defect the cooperative behavior does not survive and the system dynamics evolves to an all-D state. Surprisingly the cooperation is enhanced by stopping the growth for a wide range of b values, pointing out that the cooperation levels observed during growth are very robust. Moreover, the value of b_c appears to increase with the intensity of selection ϵ in agreement with the increase of the degree heterogeneity of the substrate network. These results highlight the phenomenological difference between playing the PD game simultaneously to the growth of the underlying network and playing on fixed static networks.

7.5.2 Effects of Randomizations on the Evolutionary Dynamics

Now, in order to gain more insight in the relation between network topology and the supported level of cooperation, we study the evolution of cooperation when network growth is stopped and we make different randomizations of both the local structure and the strategies of the nodes. In particular, in Fig. 7.8, we show the asymptotic level of cooperation when the following randomizations are made after the growth is stopped: *(i)* the structure of the EPA network is randomized by rewiring its links while preserving the degree of each node; *(ii)* the structure of the network is kept intact but the strategies of the nodes are reassigned while preserving the global fraction of cooperation (strategy randomization); and *(iii)* when the two former randomization procedures are combined.

As it can be seen from Fig. 7.8, the crucial factor for the cooperation increment during the size-fixed period of the dynamics is the structure of these EPA networks, since its randomization leads to a decrease of cooperation at levels far away from those of the original one or even of a BA SF network [2, 18]. This drop of cooperation when randomizing the structure is in good agreement with previous findings in complex topologies, specifically, for static BA networks [11, 19] (see also Sect. 5.1). On the other hand, the strategy randomization does not prevent high levels of coop-

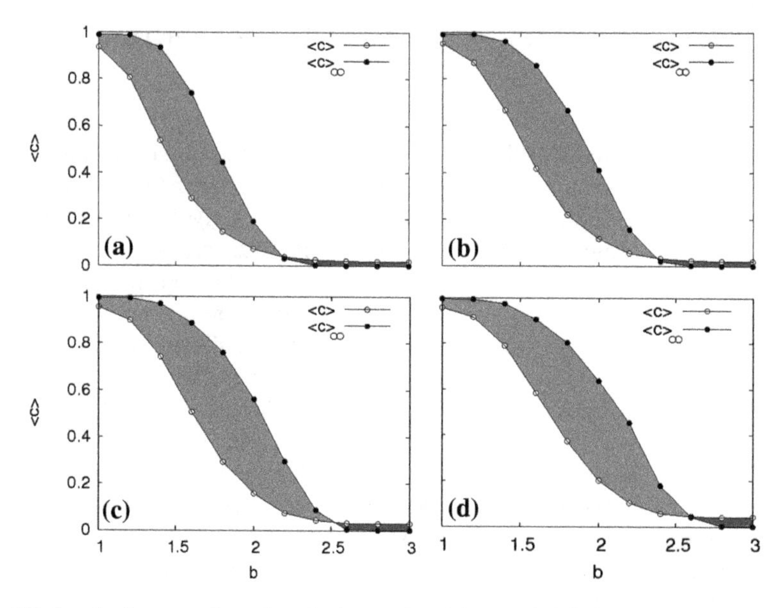

Fig. 7.7 Level of cooperation when the last node of the network is incorporated, $\langle c \rangle$, and the average fraction of cooperators observed when we let the system evolve after the network growth has ended, $\langle c \rangle_\infty$. The four panels show these measures for several values of ϵ. From top to bottom and left to right we show $\epsilon = 0.5, 0.75, 0.9$ and 0.99 (strong selection limit). The networks are made up of $N = 10^3$ nodes with average connectivity $\langle k \rangle = 4$ and $\tau_D = 10\tau_T$. Every point is the average over 10^3 realizations

eration, thus confirming that the governing factor of the network behavior is the structure arising from the co-evolutionary process. Moreover, the asymptotic level of cooperation in this case (squares in Fig. 7.8) is larger that those observed when the network is simply let to evolve without any randomization (C_∞ in Fig. 7.7). This result points out that using a random initial condition for the strategies differs strongly from starting from a configuration where degrees and strategies are correlated as a result of the EPA model (Fig. 7.3). We will come back to this point in Sect. 7.7.

7.5.3 EPA Networks as Substrates for Evolutionary Dynamics

The high levels of cooperation observed when applying a random initial configuration for the strategies to EPA networks motivate the question on whether EPA networks are best suited to support cooperative behavior than other well-known models. In order to answer this question, we consider our EPA networks when used as static substrates for the evolutionary dynamics and compare with the cases of both Barabási-Albert [18] and Erdős-Reńyi (ER) [20] graphs. To this aim, we take a particular example of our model networks, grown with $b = 2.1$ and $\epsilon = 0.99$, and run the evolutionary dynamics starting from an initial configuration with 50% coopera-

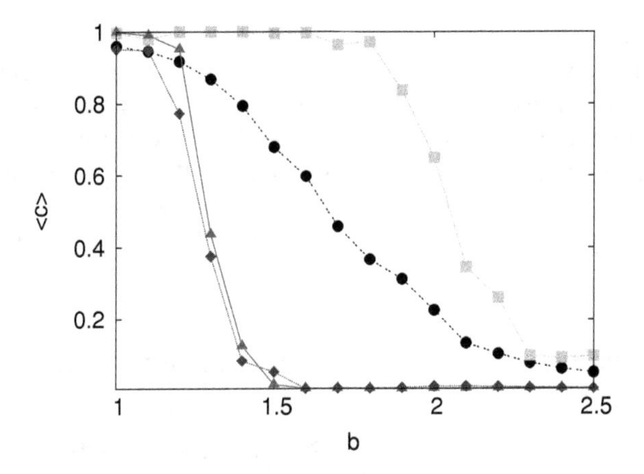

Fig. 7.8 Cooperation levels at the end of the growth process and after letting the network relax as a function of b. The original network was grown up to $N = 4 \cdot 10^3$ nodes with $\epsilon = 0.99$ and average connectivity $\langle k \rangle = 4$, and the asymptotic cooperation levels are computed 10^7 time steps afterwards. Full circles show the cooperation level when the network stops growing. The other curves show the asymptotic cooperation when the structure of the network has been randomized (*triangles*), when the strategies of the nodes have been reassigned randomly (*squares*) and with both randomizations processes (*diamonds*)

Fig. 7.9 Cooperation levels in ER, BA, and our Evolutionary Preferential Attachment network models, as a function of the temptation parameter b. The EPA network is built up using the model described in the main text for $b = 2.1$ and $\epsilon = 0.99$. All networks are made up of $N = 10^3$ nodes, with average connectivity $\langle k \rangle = 4$, and every point shown is the average over 10^3 independent realizations

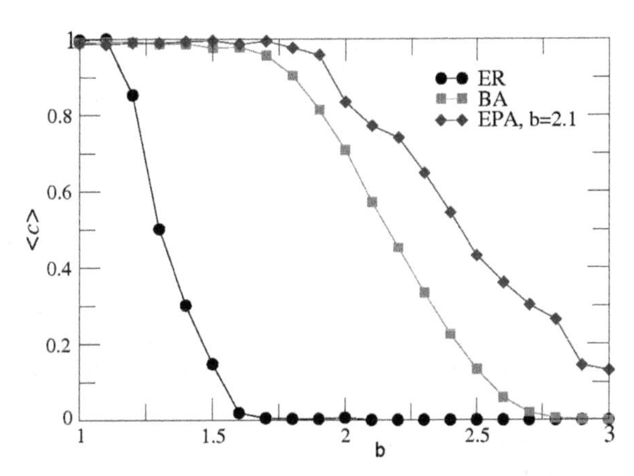

tors and defectors placed at random. The average level of cooperation as a function of the temptation to defect is represented in Fig. 7.9 together with the diagrams for BA and ER networks. Surprisingly, the plot shows that the EPA network remarkably enhances the survival of cooperation for all the values of b studied. Therefore, the attachment process followed by EPA networks seems to be more efficient than the BA preferential attachment model studied in [8, 14, 21]. Obviously, the roots of this behavior cannot be found in the degree distribution, $P(k)$, but in the high levels of clustering [22] and the disassortativeness [23] shown above.

It is worth mentioning here that we have performed an study of the asymptotic state of the system, and we have computed the fractions of pure strategist and fluctuating individuals (as we have defined them in Sect. 3.3), once the network has grown to its final size. But since they are not very novel results, we will not discuss them right now. Instead, we will show them as a comparison with the case $\tau_D = \tau_T$, in Sect. 7.8. We just confirm here the existence of the partition of the (static) EPA network into the usual sets of pure cooperators, pure defectors and fluctuating individuals.

7.6 Time Evolution of the $P_c(k)$ After Network Growth

As it has been well established before, SF topologies are able to sustain higher levels of cooperation than random structures due to the microscopical organization of the strategies [8, 14]. In particular, it has been shown that in those heterogeneous settings the hubs always play as cooperators being surrounded by a unique cluster of cooperators, while defectors cannot take advantage of high connectivity, and thus occupy medium and low degree classes. Nonetheless, in our EPA structures, we have observed (Sect. 7.3) that while the network grows, some hubs play as defectors, thus implying a very different microscopic scenario than that of static heterogeneous networks.

In this section we turn again to the situation in which the network growth has stopped (and no randomization is made) to study the roots of the increment of the asymptotic level of cooperation reported in Fig. 7.7.

To this aim we look at the temporal evolution of the probability that a node of degree k is a cooperator, $P_c(k)$, once the network growth has ceased. As we have observed in Sect. 7.3, the growth process leads to a concentration of cooperators at nodes with intermediate degree, explained from the fact that while the network is growing, newcomers join in with the same probability of being cooperators or defectors. In this situation, defectors have an evolutionary advantage as they get higher payoffs from cooperator newcomers. Although these cooperators will eventually change into defectors and stop providing payoff for the original defector, the stable source of fresh cooperator nodes entering the network compensates for this effect. However, when the growth stops while the dynamics continues, we observe that low degree nodes are rapidly taken over by cooperators, and after 10^4 time steps they are mainly cooperators. On the contrary, hubs are much more resistant to change, and even after 10^7 time steps not all of them have changed into cooperators (revealed by those values $P_c(k) = 0$ in Fig. 7.10).

The persistence of hub defectors is a very intriguing observation, in contrast with previous findings in static SF networks [8, 14, 19] (see also Chap. 3), for which hubs are always cooperators or, in other words, a defector hub is unstable. As we have widely explain in Chap. 3, this occurs because a defector sitting on a hub will rapidly convert its neighbors to defectors, which in turn leaves it with zero payoff; subsequently, if one of its neighbors turns back to cooperation, the hub will eventually follow. It seems, however, that the coupling of evolutionary game dynamics with the

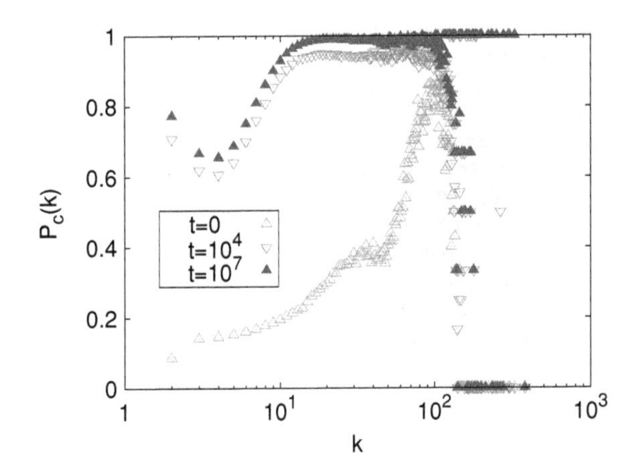

Fig. 7.10 Probability of being a cooperator as a function of the degree at the end of the Evolutionary Preferential Attachment process, 10^4 time steps later, and 10^7 time steps later, for $b = 2.2$ and $\epsilon = 0.99$

network growth leads to a structural and dynamical configuration that stabilizes the defectors on hubs. The unexpected result that Fig. 7.10 shows is that defector hubs can also be asymptotically stable once the network growth has ceased, i.e., it has become static. Indeed, we have observed in our simulations that some hubs are defectors for as long as the dynamics continues (at least, $t = 10^7$ extra time steps after finishing growing the network). However, it is important to stress that not all realizations of the process end up with defector hubs. For low values of b, this is practically never the case and almost no realizations produce defectors at the hubs. However, as b increases, the percentage of realizations where this phenomenon is observed increases rapidly.

In Sect. 7.3 we have discussed why a hub can be a defector while the network is growing: it is because it takes advantage of the newcomer flow, getting high benefits from them. Nevertheless, the surprising fact that defector hubs may have very long lives on the static regime, may be the relevant feature for the behavior of the network resulting from the growth process, and it is important to fully understand the reason for such a slow dynamics. We claim that it can be traced back to the payoff structure of the network, so in Sect. 7.7, we will analyze it in detail.

7.7 Microscopic Roots of Cooperation After Network Growth

Having identified the coexistence of cooperator and defector hubs, we next study why this configuration seems to be asymptotically stable and why the hubs are not invaded by opposite strategies. In Fig. 7.11, we present an example taken from a single realization of the process. Had we plot the results of payoffs averaged over realizations, we would not have been able to obtain this picture, because in that case payoffs are seemingly very different in the region of large degree, as a consequence of the statistical properties of our networks, in which hubs do exist but their degree

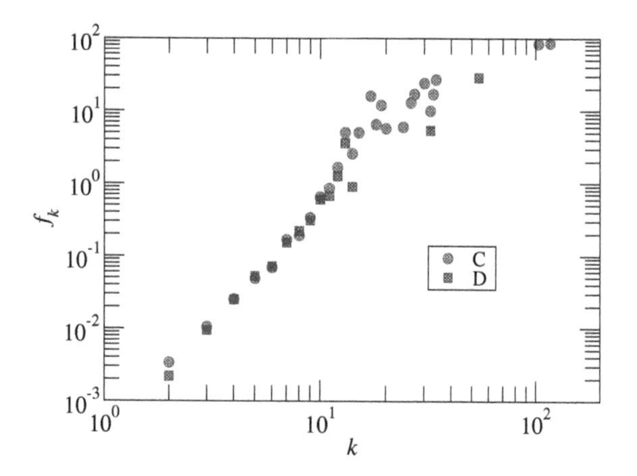

Fig. 7.11 Average payoff for cooperators and defector nodes at the end of network growth ($t = 0$) as a function of their degree, k, for a realization of the Evolutionary Preferential Attachment model with $b = 1.8$. Note that the similarity between cooperators' and defectors' payoffs implies that imitation events -or invasions- take place on a long time scale

and payoff depend on the specific realization. As can be seen, the payoff grows approximately as a power law, $f_k \sim k^{\alpha}$; however, the key point here is not this law but the fact that the payoffs for defectors and cooperators of the same degree are very similar. In view of the strategy update rule (Eq. 7.1), it becomes clear that the evolution must be very slow. Moreover, if we take into account the role of the degree in that expression, we see that hubs have a very low probability to change their strategies, whatever they may be.

Considering now the disassortative nature of the degree–degree correlations (Fig. 7.6) we can explain how these dynamical configurations can be promoted by the structure of the network. The large dissasortativity of EPA networks suggests that hubs are mostly surrounded by low degree nodes and not directly connected to other hubs. Instead, the connection with hubs is made in two steps (*i.e.* via a low degree node). This local configuration resembles that of the so-called Dipole Model [24] and 3.4, a configuration in which two hubs (not directly connected) are in contact with a large amount of common neighbors which in turn are low degree nodes. In this configuration, it can be shown analytically that the two hubs can coexist asymptotically with opposite strategies, provided that the hub playing as cooperator is in contact with an additional set of nodes playing as cooperators, for this will provide the hubs with a stable source of benefits. On the contrary, defector hubs are only connected to the set of nodes that are also in contact with the cooperator hubs. In this setting, the low degree individuals attached to both hubs experience cycles of cooperation and defection (we call them *fluctuating individuals*, because their strategies can never get fixed) due to the high payoffs obtained by the hubs. If such a local configuration for the strategies of hubs and their leaves arises, neither of the two hubs will take over the set of fluctuating individuals, nor the latter will invade the hubs as they are mainly poorly connected nodes with small payoffs.

In order to test if the grown networks exhibit local dipole-like structures, we have measured the connectivity of the neighbors of defector and cooperator hubs, which we represent in Fig. 7.12. The figure undoubtedly shows that highly connected nodes

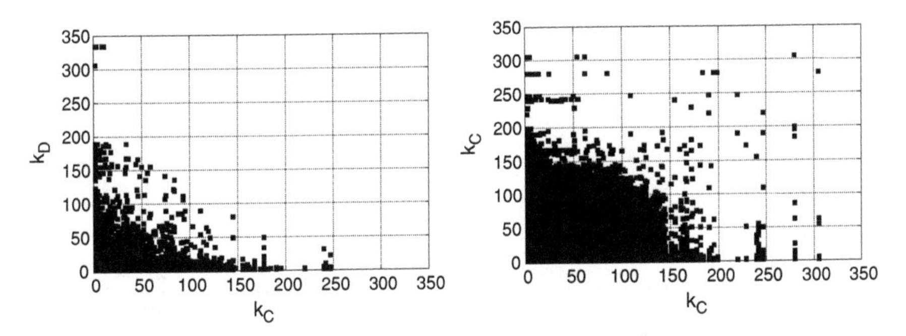

Fig. 7.12 Connectivity matrix of cooperators with defectors (*left*) and of cooperators with themselves (*right*) for a single realization of the process. The element (i, j) is set to 1 (*black square* in the figure) when a link between a defector (cooperator) of degree i and a cooperator (cooperator) of degree j exists, respectively

playing as defectors are mainly connected to poorly connected cooperators (acting as the set of fluctuating strategists), whereas cooperator hubs are connected to each other and also to a significant fraction of lowly connected nodes. This fully confirms that, in contrast to all previous results, there is a structure allowing the resilience of defector hubs, and moreover, it gives rise to a situation quite similar to that described by the Dipole Model.

7.8 Other τ_D/τ_T Time Relations

During this whole chapter, we have always worked with a time relation between the dynamics and the growth of the network equal to $\tau_D = 10\tau_T$, meaning that the network grows in ten nodes at the time, and then one single round of the dynamics takes place. We have studied the degree distributions that can arise from this Evolutionary Preferential Attachment mechanism, as well as the levels of cooperation, comparing them with some well-known cases, such as BA scale-free or ER random static networks. Nevertheless, it is interesting to explore the behavior of the system for other time ratios. Specifically, now we will explore briefly the case when both time scales are exactly the same $\tau_D = \tau_T$, i.e., starting with a small core of nodes fully connected, we add a new node at a time and then we make the system play one round of the game. We will compare the results with the $\tau_D = 10\tau_T$ scenario.

Thus, in Fig. 7.13 we show some degree distributions obtained for this particular time relation, and as we can see, there are some qualitative differences between this case and the one with $\tau_D = 10\tau_T$ one (see Fig. 7.1 to compare them). First of all, if we look at the two upper panels, we can see that the dependence of $P(k)$ with ϵ and for a fixed value of the temptation to defect is less clear in this case, while it was obvious and very gradual for the $\tau_D = 10\tau_T$ scenario. Also, when $\epsilon = 0.99$, the networks that arise from the process have very fat-tailed degree distributions (even more so for high values of the parameter b), which means that there are 'super-hubs'

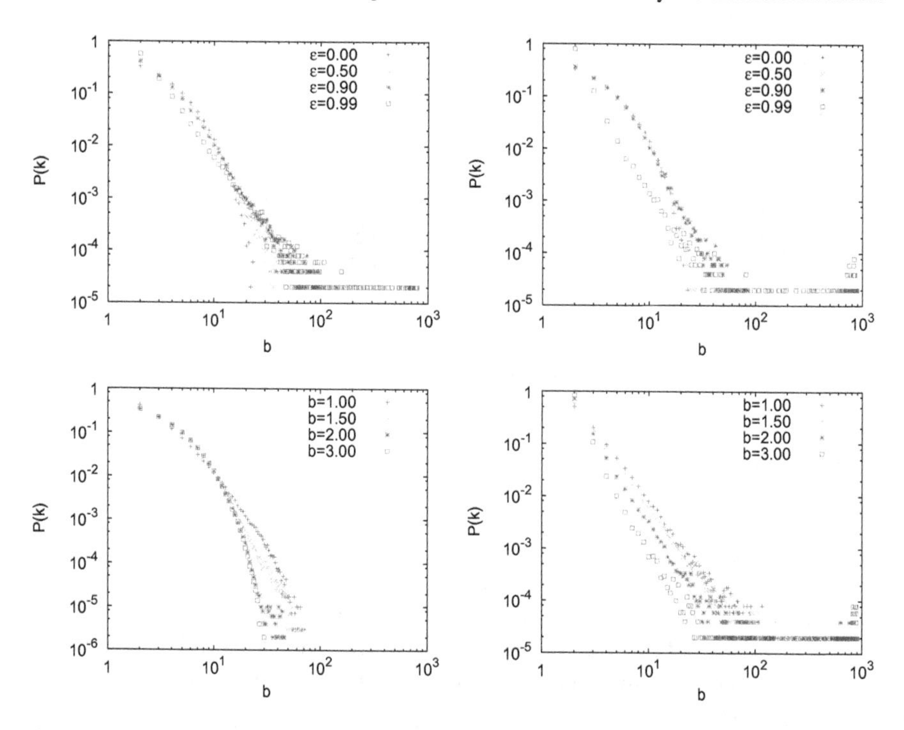

Fig. 7.13 Degree distribution for fixed values of $b = 1.5$ (*Top left*) and $b = 2.5$ (*Top right*), and fixed values of $\epsilon = 0.3$ (*Bottom left*) and $\epsilon = 0.99$ (*Bottom right*). The networks are made up of $N = 10^3$ nodes, with average connectivity $\langle k \rangle = 4$, and $\tau_D = \tau_T$. Every point is the average of 300 different realizations

present in the system, which were not there in the previous case. On the other hand, there is a more pronounced dependence on the parameter b for a fixed value of ϵ (bottom panels of Fig. 7.13), while for the $\tau_D = 10\tau_T$ case, the degree distributions were almost b-independent.

In order to characterize better the behavior of the system when the time relation is $\tau_D = \tau_T$, we also need to look at the level of cooperation, comparing the $\langle c \rangle(b)$ curves, as well as the fractions of pure strategist and fluctuating individuals for several cases. But first of all, we need to point out an important difference between the present scenario and the one studied in previous sections. In the situation with $\tau_D = 10\tau_T$, we observed that the final state of the system was, in general, fluctuating around a well-defined value of cooperation, so the interpretation of the magnitude $\langle c \rangle$ was the fraction of cooperation present in the network in the stationary state. Nonetheless, for the case we are studying now, the situation is different, since the system always reaches an all-C or an all-D state. Thus, one should interpret $\langle c \rangle$ as the fraction of realizations for which the system ends up in an all-C state. Now, as we can see in Figs. 7.14 and 7.15 for both extreme values of ϵ, the weak and strong selection limits, the average level of cooperation is remarkably lower for the case of $\tau_D = \tau_T$. This fact can be understood as follows: if we start with a small core of

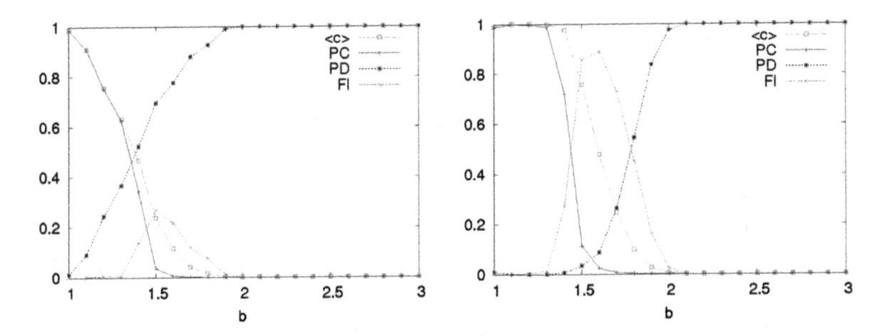

Fig. 7.14 Average level of cooperation and fractions of pure strategists and fluctuating individuals as a function of b, for $\tau_D = \tau_T$ (*Left*) and $\tau_D = 10\tau_T$ (*Right*), both for $\epsilon = 0.0$ (weak selection limit). The networks are made up of $N = 10^3$ nodes, with average connectivity $\langle k \rangle = 4$. Every point is the average of 300 independent realizations

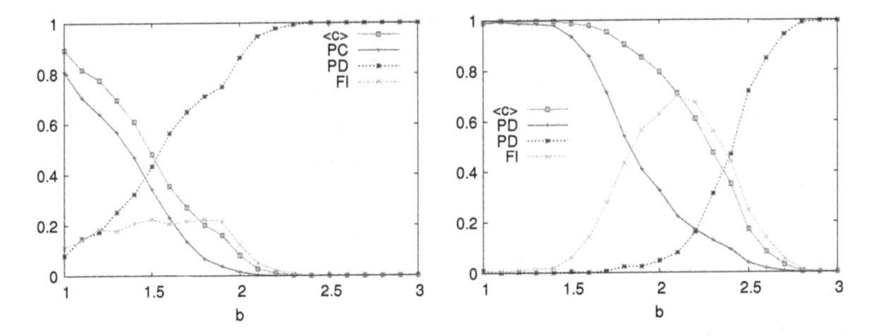

Fig. 7.15 Average level of cooperation and fractions of pure strategists and fluctuating individuals as a function of b, for $\tau_D = \tau_T$ (**Left**) and $\tau_D = 10\tau_T$ (**Right**), both for $\epsilon = 0.99$ (strong selection limit). The networks are made up of $N = 10^3$ nodes, with average connectivity $\langle k \rangle = 4$. Every point is the average of 300 independent realizations

nodes fully connected, and the networks grows very slowly (since the time relation is now $\tau_D = \tau_T$), the situation is in many ways similar to a well-mixed scenario, where it has been proved that the cooperation cannot survive [25–28] (see Sect. 2.2.2). On the contrary, if the network grows faster (for example, when the relation $\tau_D = 10\tau_T$ is fulfilled), the cooperation has better chances to survive, due to the structure of the graph. We can also notice that the level of fluctuating individuals is lower for the $\tau_D = \tau_T$ situation, since the pure defectors start invading the network much earlier, it is to say, for much lower values of the temptation to defect.

We have also tried other time relations, such as $10\tau_D = \tau_T$, it is, a new node is added, and then the system plays 10 rounds of the game. Obviously, in this case we have found the same well-mixed effect than in the $\tau_D = \tau_T$ but enhanced: the level of cooperation drops even more, because this new scenario promotes cooperation even less than the previous one.

7.9 Conclusions

In this chapter we have presented a model in which the rules governing the formation of the network are linked to the dynamics of its components. The model provides an evolutionary explanation for the origin of the two most common types of networks found in natural systems. Thus, when the selection pressure is weak, homogeneous networks arise, whereas strong selection pressure gives rise to scale-free networks. A remarkable fact is that the proposed evolution rule gives rise to complex networks that share many topological features with those measured in real systems, such as the power law dependence of the clustering coefficient with the degree of the nodes. Interestingly, our results make it clear that the microscopic dynamical organization of strategists in evolutionarily grown networks is very different from the case in which the population evolves on static networks. Namely, there can be hubs playing as defectors during network growth, while cooperators occupy mainly the middle classes. It is worth stressing that the level of cooperation during network growth reaches the highest values for the strong selection limit in which the newcomers launch their links to the fittest elements in the system.

Furthermore, the generated networks are robust in the sense that after the growth process stops, the dynamical behavior keeps its character. Moreover, we have shown that for most cases the cooperative behavior arising in these networks exhibits a great resilience, in the sense that it does not decrease for a wide range of parameters upon stopping the growth process, and, in most cases, it even displays a large increase of the cooperation level. We have also shown that the non-trivial topological patterns of EPA networks are the roots for such enhancement of the cooperation. In particular, we have shown that rewiring the links while keeping the degree distribution (thus destroying any kind of correlations between nodes) yields a dramatic decrease of the levels of cooperation. On the other hand, a randomization of the strategies does not affect the asymptotic levels of cooperation. Therefore, the ability of EPA networks to promote the resilience of cooperation is rooted in the correlations created during network formation via the co-evolution with the evolutionary dynamics.

Finally, maybe the most important difference we have found between the networks grown with our model and the static SF case, is the dynamic stabilization of defectors on hubs. We have shown that these defector hubs can be extremely long-lived due to the similarity of payoffs between cooperators and defectors arising from the co-evolutionary process. Moreover, we have been able to link the payoff distribution to the network structure. In particular, we show that the disassortative nature of EPA networks together with the formation of local dipole-like structures [24] (and see also Sect. 3.4) during network growth is responsible for the fixation of defection in hubs.

References

1. M. Newman, SIAM Review **45**, 167 (2003).
2. S. Boccaletti, V. Latora, Y. Moreno, M. Chavez, and D. Hwang, Phys. Rep. **424**, 175 (2006).
3. G. Bianconi and A. L. Barabási, Europhys. Lett. **54**, 436 (2001).
4. G. Caldarelli, A. Capocci, P. D. L. Rios, and M. A. M. noz, Phys. Rev. Lett. **89**, 258702 (2002).
5. A. Rapoport and A. M. Chammah, *Prisoner's Dilemma.* (Univ. of Michigan Press, Ann Arbor, 1965).
6. K. Lindgren and M. Nordahl, Physica D **75**, 292 (1994).
7. M. A. Nowak and R. M. May, Nature **359**, 826 (1992).
8. F. C. Santos and J. M. Pacheco, Phys. Rev. Lett. **95**, 098104 (2005).
9. H. Gintis, *Game theory evolving.* (Princeton University Press, Princeton, NJ, 2000).
10. C. Hauert and M. Doebeli, Nature *428*, 643 (2004).
11. F. C. Santos, F. J. Rodrigues, and J. M. Pacheco, Proc. Biol. Sci. *273*, 51 (2006).
12. J. Hofbauer and K. Sigmund, *Evolutionary games and population dy- namics.* (Cambridge University Press, Cambridge, UK, 1998).
13. J. Hofbauer and K. Sigmund, Bull. Am. Math. Soc. **40**, 479 (2003).
14. J. Gómez-Gardeñes, M. Campillo, L. M. Floría, and Y. Moreno, Phys. Rev. Lett. *98*, 108103 (2007).
15. M. Nowak, A. Sasaki, C. Taylor, and D. Fudenberg, Nature *428*, 646 (2004).
16. M. Nowak, Science **314**, 1560 (2006).
17. E. Lieberman, C. Hauert, and M. A. Nowak, Nature **433**, 312 (2005).
18. A. Barabási and R. Albert, Science **286**, 509 (1999).
19. F. C. Santos and J. M. Pacheco, J. Evol. Biol. **19**, 726 (2006).
20. P. Erdos and A. Renyi, Publicationes Mathematicae Debrecen **6**, 290 (1959).
21. F. C. Santos, J. M. Pacheco, and T. Lenaerts, Proc. Natl. Acad. Sci. USA **103**, 3490 (2006).
22. S. Assenza, J. Gómez-Gardeñes, and V. Latora, Phys. Rev. E **78**, 017101 (2008).
23. A. Pusch, S. Weber, and M. Porto, Phys. Rev. E **77**, 036120 (2008).
24. L. M. Floría, C. Gracia-Lázaro, J. Gómez-Gardeñes, and Y. Moreno, Phys. Rev. E **79**, 026106 (2009).
25. R. Axelrod, *The complexity of cooperation: agent-based models of com- petition and collaboration.* (Princeton University Press., Princeton, NJ, 1997).
26. M. Nowak, *Evolutionary dynamics: exploring the equations of life.* (Harvard University Press., Cambridge, MA, 2006).
27. M. Nowak and K. Sigmund, *Games on Grids, in: The Geometry of Ecological Interactions.* (Cambridge University Press, Cambridge, UK, 2000).
28. R. Axelrod and W. Hamilton, Science **211**, 1390 (1981).

Chapter 8
Complex Networks from Other Dynamic-Dependent Attachment Rules

In this chapter, we will continue exploring the issue of the entanglement between the growth of a complex structure and the dynamics that is taking place on top of it simultaneously, in such a way that the outcome of the game, meaning the benefits the nodes get out of the interaction, will affect the probability of the existing nodes to attract links from newcomers. So we will work with a model similar to the one introduced in Chap. 7, but with two important differences: on the one hand, the dependence of the probability of attachment will be exponential with the fitness of the nodes, instead of linear. On the other hand, we will also modify the imitation rule to a Fermi-like function, instead of using a Replicator-like probability, so irrational changes of strategy will be allowed now, meaning that a node can imitate a neighbor whose payoff is lower than its own.

The approach we will take here will be a little different too. Since this model has one more parameter than the one exposed in Chap. 7, instead of presenting it at once, considering simultaneously all the effects, we will study first a case where the dynamics has no effect on the growth, just to separate the two contributions, and then we will take the dynamics into consideration, too.

In the model we presented here, new individuals establish connections to the existing individuals, and the newcomers can either connect to m arbitrary individuals or preferentially attach to those that have been successful players in the past, depending on the values of the corresponding parameter. Success is based on the cumulated payoff π from a round of an evolutionary game, which each individual plays with all its neighbors on the network. Although for the model itself we do not need to specify the kind of game or the number of strategies, we will use the two-strategy Prisoner's Dilemma, as in Chap. 7. However, the formulation of the game, it is to say, the values of the coefficients of the payoff matrix, will be different. We will use the cost-benefit ratio approach, like we did in Chap. 6.

J. Poncela Casasnovas, *Evolutionary Games in Complex Topologies*, Springer Theses,
DOI: 10.1007/978-3-642-30117-9_8, © Springer-Verlag Berlin Heidelberg 2012

8.1 The Model

We start from a small complete network of m_0 individuals with one strategy. Subsequently, new individuals arrive and form connections to existing individuals. Evolutionary dynamics proceeds in the following way: At each time step, every individual j plays with all its neighbors and obtains an accumulated payoff π_j. All players choose then synchronously between their old strategy and the strategy of a randomly selected neighbor. In this way, player j will adopt the strategy of its neighbor i with probability [1–5]:

$$T_{j \to i} = \frac{1}{1 + e^{\beta(\pi_j - \pi_i)}} \tag{8.1}$$

where β is the intensity of selection. Obviously, with probability $(1 - T_{j \to i})$, node j will stick to its old strategy. This updating rule is usually called Fermi rule, since it is based on the Fermi distribution function from Statistical Mechanics. The parameter β, which in Physics means inverse of temperature, can be here also interpreted as noise associated with errors in the decision making process [6]. Thus, depending on the value of this parameter, we can have now different limiting situations:

- For $\beta \ll 1$, selection is weak and the game is only a linear correction to random strategy choice, it is to say, a random drift process.
- For strong selection, $\beta \to \infty$, node j will always adopt a better strategy and it will never adopt a worse strategy (imitation dynamics).

It is important to stress that, by using this strategy updating rule, we allow individuals to be irrational, in the sense that they can adopt a strategy that performs worse than its own current one.

Every τ time steps, a new individual with a random strategy is added to the system. It means that when $\tau \ll 1$, several nodes are added before one round of the dynamics takes place on the system, and when $\tau \gg 1$, the network grows very slowly and the game dynamics can bring the system close to equilibrium before a new node is added. The new individual establishes m links to preexisting nodes, which are chosen preferentially according to their performance in the game in the last time step. Node j is chosen as game partner with probability:

$$p_j = \frac{e^{+\alpha \pi_j}}{\sum_{l=1}^{N(t)} e^{+\alpha \pi_l}} \tag{8.2}$$

where $N(t)$ is the number of nodes that already exist when the new node is added at time t. The remaining $m - 1$ links are added in the same way, excluding double links, as usual. Again, one should realize that different cases are possible, depending on the value of the parameter α:

- For $\alpha = 0$, the newcomer attaches to a randomly chosen existing node.
- For small α, attachment is approximately linear with payoff.

- For high α, the newcomers will make connections to only a small set of nodes with the high payoffs.
- In the limit $\alpha \to \infty$, all newcomers will always attach to the m most successful players.

Besides, since m links and a single node are added at each τ time step, the average degree of the network at a given moment is:

$$\langle k \rangle (t) = \frac{m_0(m_0 - 1)\frac{1}{2} + m\frac{t}{\tau}}{m_0 + \frac{t}{\tau}} \tag{8.3}$$

where t is the number of time steps that has passed. Throughout this chapter, we will use $m = 2$ (therefore, $\langle k \rangle = 4$) and $m_0 = 3$.

8.1.1 A Simplification of the Model

As we have mentioned previously, in order to fully understand this model and the different contributions each feature makes to the final outcome, we want to focus on the simplest case, in which each interaction leads to the same payoff, which we set to one. Or in other words, it would correspond to a game whose entries of the payoff matrix were all equal: it does not make any difference which strategy you or your opponent may choose. Then, the payoffs π_j are just the number of interactions an individual has, i.e. the degree k_j of the node (note that normalizing by the degree of the node would essentially wash out the effect of the topology at this point [7, 8]).

Thus, evolutionary dynamics of strategies has no consequences and thus, the topology is independent of β. This allows us to discuss the growth dynamics without any complications arising from the dynamics of strategies. We have several simple limiting cases:

- For $\alpha = 0$, the newcomer attaches at random to any pre-existing node. This leads to a network in which the probability that a node has k links decays exponentially, similar to ER networks. In this case, topology is independent of strategies for all intensities of selection β, even when individuals play different strategies leading to different payoffs. Nonetheless, whenever $\alpha > 0$, there is an interplay between topological dynamics and strategy dynamics.
- For $\alpha \ll 1$, we can linearize the probability of attachment p_j, and we obtain:

$$p_j = \frac{\alpha^{-1} + k_j}{\sum_{k=1}^{N} (\alpha^{-1} + k_k)}. \tag{8.4}$$

  Thus, we recover the linear preferential attachment model introduced by Dorogovtsev et al [9]. When strategies differ in their payoffs, then not only the degree, but also the strategy of the nodes and their neighbors will influence the attachment probability.

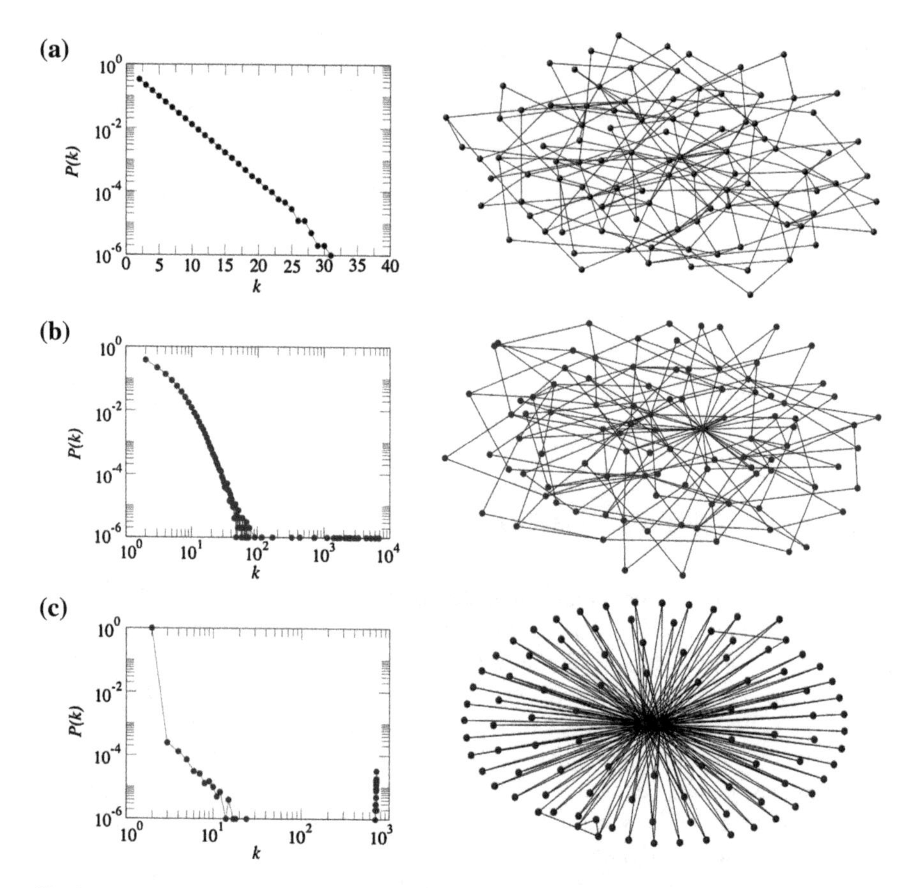

Fig. 8.1 Networks for a game in which both strategies have identical payoffs, such that the payoff is given by the degree of a node. The *left* hand side shows the degree distributions of networks of size $N = 10^4$, while the *right* hand side shows snapshots of networks of $N = 100$ nodes. **a** For $\alpha = 0.0$, the degree distribution decays exponentially. **b** For $\alpha = 0.1$, some highly connected nodes appear in the network and the degree distribution begins to resemble a power-law. **c** Already for $\alpha = 1.0$, the vast majority of nodes (>99.9 %) has only two links. In addition, $\langle k \rangle = 2\text{m} = 4$ of the $m_0 = 3$ initial nodes are connected to almost all other nodes. Degree distributions are obtained from an average over 10^2 networks of size $N = 10^4$. Note that the x-axis is linear in **a**, but logarithmic in **b** and **c**

- When α is large, we will typically observe a network in which m of the m_0 nodes of the initial complete network will be connected to almost all nodes that have been added during the growth stage. The emergence of these super-hubs is due to the nonlinearity in Eq. 8.2.

Examples for the network structures in these limiting cases are given in Fig. 8.1. As it is shown, for $\alpha = 0$, random networks are generated. On the other hand, when α increases, some degree of heterogeneity appears in the resulting structure, whereas for $\alpha = 1$, the probability of attachment is so strongly dependent of the connectivity, that it exclusively benefits m among the m_0 initial nodes, that become super-hubs, and so the model always gives rise to star-like structures.

Next, we will go back to evolutionary games in which the payoff per interaction is no longer constant, but depends on the strategies of the two interacting individuals. In general, such an interplay of evolutionary dynamics of the strategies and the payoff-preferential attachment will change the structure of the network.

8.2 Degree Distribution

After this brief study of a simplified version, let's now address the whole model again. The dynamics we will consider here is once again the Prisoner's Dilemma [10–12], where the two players can choose between two possible strategies: cooperation (C) and defection (D). But as we have mentioned before, in this case, the values of the coefficients of the payoff matrix will be different from those we used mainly in previous chapters, although the relative ordering of them must remain the same. Namely, the parameter that characterizes how expensive cooperation is, compared with defection, will be the ratio b/c, instead of using the temptation to defect b. In this way, we will consider that there is a cost c for cooperation, whereas a cooperative act from an interaction partner leads to a benefit b ($> c$). Thus, the lower the value of b/c is, the more expensive the cooperation is. The payoff matrix of the game can be written as:

$$\begin{array}{c} \\ C \\ D \end{array}\!\!\begin{array}{c} \begin{matrix} C & D \end{matrix} \\ \begin{pmatrix} b-c & -c \\ b & 0 \end{pmatrix} \end{array} \sim \begin{array}{c} \\ C \\ D \end{array}\!\!\begin{array}{c} \begin{matrix} C & D \end{matrix} \\ \begin{pmatrix} b/c-1 & -1 \\ b/c & 0 \end{pmatrix} \end{array} \tag{8.5}$$

No matter what the opponent does, defection always leads to a higher payoff, because $b > b-c$ and $0 > -c$, thus selfish, rational players should defect. Similarly, if the payoff determines reproductive fitness, evolution will lead to the spread of defection. However, the payoff for mutual defection is smaller than the payoff for mutual cooperation ($b - c > 0$) and thus players face a dilemma. As we discussed in previous chapters, one way to resolve it is to consider structured populations in which players only interact with their neighbors [13]. Here, we follow this line of research and consider in addition growing populations, as discussed above.

Since there is an interaction between strategy dynamics and network growth, the topology of the system will obviously change under selection. So, in Fig. 8.2, we show how it changes with the benefit to cost ratio b/c, the intensity of selection β and the attachment parameter α for the particular dynamics of the Prisoner's Dilemma game. From Fig. 8.2, it is clear that the influence of the game on the degree distribution is relatively weak, for small degrees a clear difference is only found for large α and small b/c. The distribution of the relatively few nodes with many connections, however, is more sensitive to changing either b/c or β. Moreover, as we have already learned from the simplified version of the model in Sect. 8.1.1, for a

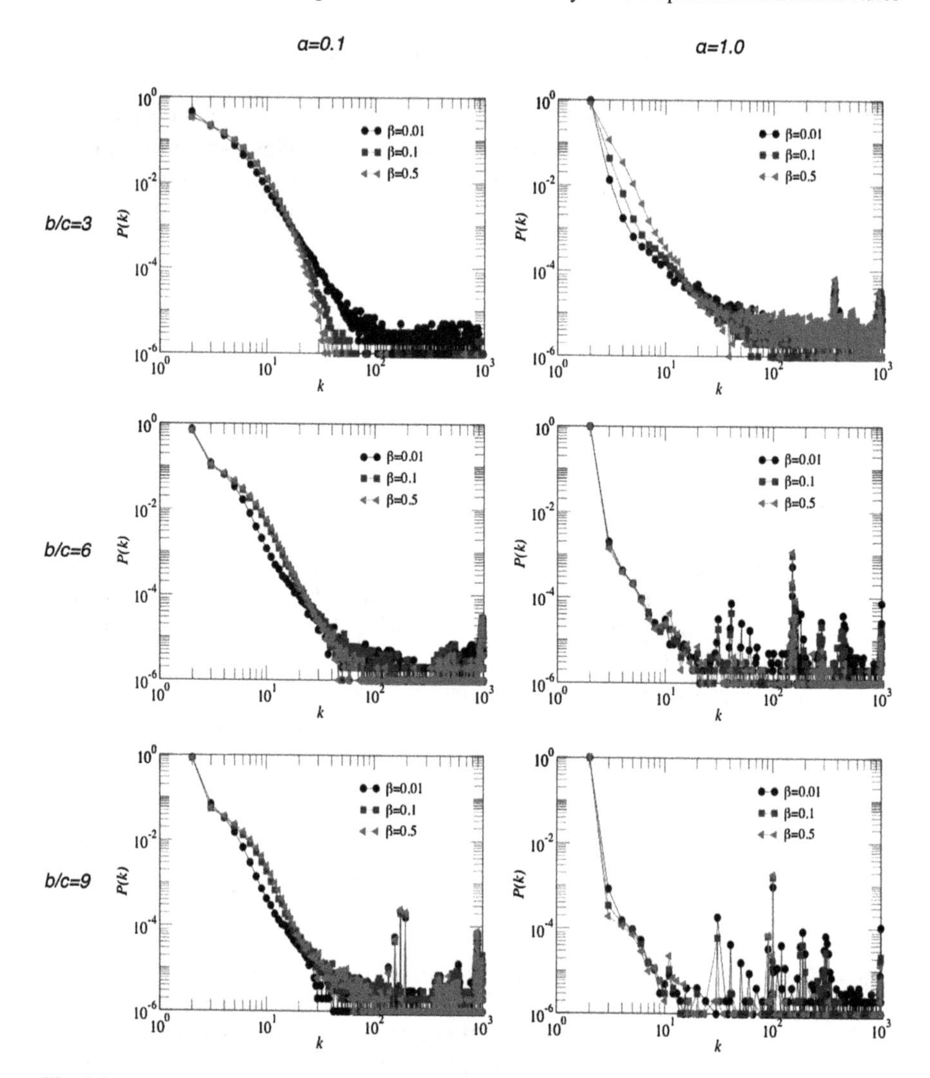

Fig. 8.2 Impact of the game dynamics on the degree distribution at the end of network growth. *Left* column corresponds to $\alpha = 0.1$, while the *right* one is for $\alpha = 1$. The networks are made of $N = 10^3$ nodes, with average connectivity $\langle k \rangle = 2m = 4$, $m_0 = 3$, and $\tau = 0.1$. All values are obtained from the average of 10^3 different realizations

value $\alpha = 1$ we have structures where super-hubs are present, regardless of the values of the other two parameters of the system, b/c and β. On the other hand, for more moderate values of α, we can observe some differences in the topologies arising from the model, depending on the values of the two other mentioned parameters. Thus, for a fixed value of the ratio benefit-cost, some different degree distributions appear, depending on β. We can also say that, in general, almost all structures obtained have

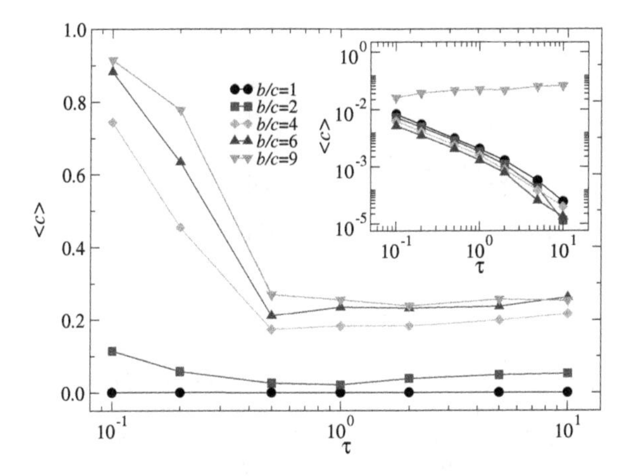

Fig. 8.3 The average level of cooperation under strong selection ($\beta = 1$) and $\alpha = 1$, depending on the time scale of attachment, τ. Cooperation benefits most from small values of τ, i.e. when many new nodes are added before players update their strategies. For random attachment ($\alpha = 0$, **inset**) cooperation does not emerge, only for high benefit to cost ratios a few cooperators prevail. The networks are made of $N = 10^3$ nodes, with average connectivity $\langle k \rangle = 2m = 4$, $m_0 = 3$, and all values are obtained from the average of 10^2 different realizations

fat-tailed $P(k)$. We can see that there is not a very important dependence of the degree distribution with b/c, which was also the case of the model presented in Chap. 7.

8.3 Average Level of Cooperation as a Function of the Parameters of the System

Typically, we are interested in the promotion of cooperation on different network structures, so Fig. 8.3 shows the average level of cooperation for strong selection as a function of τ and for several fixed values of the ratio b/c. It turns out that payoff preferential attachment increases the level of cooperation in the system significantly compared to random attachment. We want to point out here that, although we do not show it, this effect is also present for weak selection, but less pronounced. On the other hand, we observe that cooperation gets higher levels for small values of τ, i.e. when many nodes are added before dynamics takes place and strategies are changed (which is in good agreement with the results obtained in Chap. 7, where we showed that the equivalent time relation $\tau_D = 10\tau_T$ promotes cooperation much more than when $\tau_D = \tau_T$). Indeed, this particular choice for the time ratio puts the system further from equilibrium, whereas the case of large τ means that strategies have been equilibrated at least locally before the next new individual with a random strategy is added to the system. Note that for τ larger than a certain value ($\tau \lesssim 1$), cooperation levels become independent of τ, which points out that playing just once

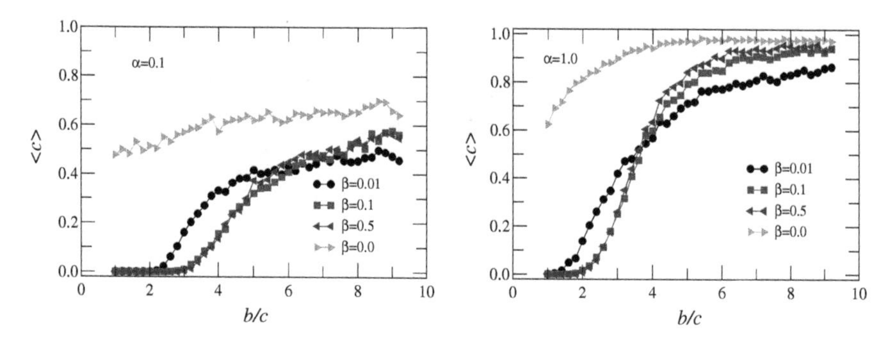

Fig. 8.4 The average level of cooperation, $\langle c\rangle$, 10^4 time steps after the network stops growing. For $\alpha = 0.1$ (*Left*) the level of cooperation exceeds 50 % only for very high benefit-to-cost ratios b/c. For $\alpha = 1.0$ (*Right*), the abundance of cooperators is significantly higher. Even for neutral strategy dynamics ($\beta = 0$), payoff preferential attachment can lead to high levels of cooperation in this case. The networks are made of $N = 10^3$ nodes, with average connectivity $\langle k\rangle = 2m = 4$, $m_0 = 3$, and $\tau = 0.1$. All values are obtained from the average of 10^3 different realizations

after a given number of new players have been incorporated is enough to reach a dynamical equilibrium.

8.4 Average Level of Cooperation After the Growth has Finished

Now, we intend to focus on analyzing the level of cooperation the system achieves once the growth has finished, it is to say, when the individuals of the network just play the game, but no new nodes are added anymore. As in most structured populations, cooperators are disadvantageous in the Prisoner's Dilemma in well-mixed population, but they can benefit from the spatial structure. Of course, this effect is larger when cooperation becomes more profitable, i.e. when the benefit to cost ratio b/c increases. It turns out that for weak payoff preferential attachment (small α), the promotion of cooperation is relatively weak and levels of cooperation beyond 50 % are only reached when cooperation is very profitable (see Fig. 8.4). However, when the probability to attach to the most successful nodes becomes large (large α), then the average fraction of cooperators becomes larger, approaching one when the benefit to cost ratio b/c is large.

Interestingly, for small b/c ratios, the abundance of cooperators decreases with increasing β, whereas it increases with the intensity of selection for large b/c ratios. The existence of a threshold for intermediate b/c can be illustrated as follows for large α: assume that we start from m_0 fully connected cooperator nodes. For $\tau < 1$, we add $1/\tau$ nodes with $m = 2$ links, half of which are defectors and half cooperators, on average. All new players interact only with the initial cooperator nodes, such that an initial cooperator will on average obtain $\frac{m}{m_0\tau}$ new links. Thus, the payoff of a new defector is mb. The average payoff of an initial cooperator is $(b - c)(m_0 - 1 +$

$\frac{1}{2}\frac{m}{m_0\tau}) - c\frac{1}{2}\frac{m}{m_0\tau}$. Both payoffs are identical for

$$\frac{b}{c} = \frac{\frac{1}{\tau} + \frac{m_0(m_0-1)}{m}}{\frac{1}{2\tau} - m_0 + \frac{m_0(m_0-1)}{m}}. \tag{8.6}$$

For large values of b/c, cooperators will dominate in the very beginning of network growth. The threshold increases with τ and decreases with m_0: the larger the initial cooperator cluster and the more nodes are added before strategies are updated, the easier it is for cooperation to spread initially. This argument shows qualitatively that a crossover in the abundance of cooperators should exist, and therefore that above a certain threshold, it is easier for cooperation to spread. Only in the very beginning of network growth, this argument will hold quantitatively.

In general, the average level of cooperation can be based on two very different scenarios: either it is the fraction of realizations of the process that ultimately ends in full cooperation, or it is the average abundance of cooperators in a network in which both cooperators and defectors are present. This also happened in the model we presented in Chap. 7: when the time relation was $\tau_D = 10\tau_T$, the average level of cooperation $\langle c \rangle$ must be interpreted as the fraction of cooperators present in the system in the stationary state, whereas for $\tau_D = \tau_T$, the whole network always ends up in a state all-C or all-D, so $\langle c \rangle$ means the fraction of realizations for which the system achieves the all-C state.

For any finite intensity of selection β, we have $T_{j \to i} > 0$, regardless of the payoffs. Thus, after growth has stopped, our dynamics describes a Markov chain with two absorbing states in which all players follow one of the two strategies. Therefore, ultimately one of the two strategies will go extinct, in contrast to evolutionary processes that do not allow disadvantageous strategies to spread. In other words, using this model, the systems will always end up whether on an all-C or on an all-D state. Nonetheless, it is important to remark that the time to extinction can become very large, in particular when the intensity of selection is high or the population size is large [4, 5, 14].

8.5 Probability of Fixation

Now, we want to analyze this issue numerically, and in order to do that, we compute the probability that fixation (for either cooperation or defection) occurs within 10^4 time steps after the network has stopped growing, during which only the dynamics takes place on the system, but no new nodes are added (see Fig. 8.5). For small α, the results follow the intuition from well-mixed populations: Fixation within this time is more likely if the intensity of selection is weaker. With increasing benefit to cost ratio, fixation times increase, so fixation within the first 10^4 time steps becomes less and less likely. For large α, however, fixation is faster for strong selection (large β) for a wide range of parameters. Only when the b/c ratio is very high, fixation times are very

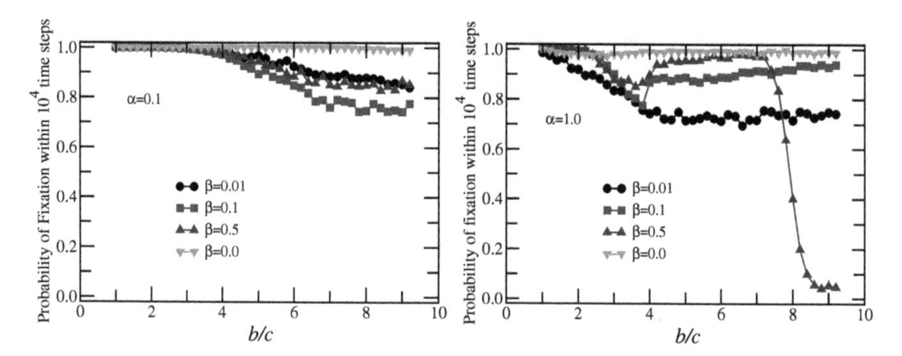

Fig. 8.5 The probability of fixation for one strategy within 10^4 time steps after growth has stopped as a function of the attachment parameter: (*Left*) $\alpha = 0.1$ and (*Right*) $\alpha = 1$, for different intensities of selection β. The networks are made up of $N = 10^3$ nodes, with average connectivity $\langle k \rangle = 2m = 4$, $m_0 = 3$ and $\tau = 0.1$. Every point is the average over 10^3 independent realizations

large under strong selection. This is based on the peculiar structure of the network obtained for large α. In addition, we observe an area in Fig. 8.5 where the fixation time increases slightly before it decreases again, i.e. the probability for fixation in the first 10^4 time steps has a minimum. Interestingly, this occurs for the range of b/c ratios where the average levels of cooperation intersect at 50 % for the different intensities of selection. In this parameter region, neither cooperators nor defectors are clearly favored. Thus, both of them spread initially. When the abundance of both strategies is approximately the same in the beginning, then it will be more difficult to completely wipe out one strategy later. Thus, the increased time of fixation in the parameter region where the abundance of cooperation becomes 50% makes intuitive sense.

8.6 Level of Cooperation After Re-Initializing the Strategies

Finally, we want to focus on studying what happens when the network stops growing: Does cooperation benefit from the growth or only from the topology? Typically, one would expect that defectors profit from the growth, because there is a steady flow of new cooperators that they can potentially exploit. Thus, cooperation should increase if the game dynamics continues on the fully grown, static network (in fact, this was the result we obtained in Chap. 7). In contrast to that case, here we have changed the game dynamics in such a way that individuals sometimes can also adopt a worse strategy (irrational changes). It has been shown in previous works that this apparently small change can significantly decrease the level of cooperation [15]. The overall level of cooperation drops significantly and is only higher than 50% if cooperation is very profitable. Indeed, we have found that with this model, the level of cooperation now decays once the network no longer grows (see Fig. 8.6). This means that in the

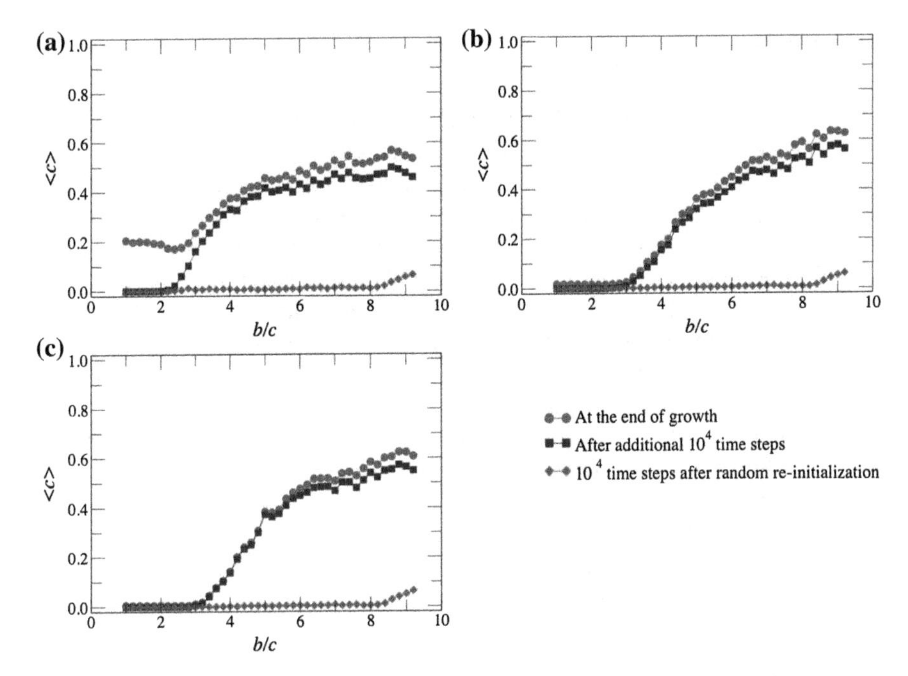

Fig. 8.6 The average level of cooperation in three cases: once the network is fully grown (*circles*), after the game dynamics has proceeded 10^4 additional steps beyond the growth phase of the network (*squares*), and 10^4 time steps after the fully grown network has been re-initialized with random strategies (*diamonds*). The intensity of selection is (**a**) $\beta = 0.01$, (**b**) $\beta = 0.1$, and (**c**) $\beta = 0.5$, respectively. The networks are made up of $N = 10^3$ nodes, with $m_0 = 3$, average connectivity $\langle k \rangle = 2\text{m} = 4$ and $\tau = 0.1$. Every point is the average over 10^2 different realizations, and $\alpha = 0.1$ in all cases

current case, cooperators, not defectors, benefit from the continuous supply of new players, so when the structure stops growing, they stop getting such high benefits, and their proportion in the system drops a little.

Thus, it makes sense to ask whether the topologies that are obtained from the network growth are cooperation promoters at all. This can be tested, as we did in Sect. 7.5, by taking the fully grown structure as a static substrate, and run the game dynamics on that fixed topology with initially random strategies, 50% cooperators and 50 % defectors. Interestingly enough, this does not lead to any significant enhancement of the level of cooperation, on the contrary, cooperators almost disappear from the system after 10^4 steps of the dynamics, once the re-initialization has been made (see Fig. 8.6). Thus, our model of network growth based on payoff preferential attachment itself leads to comparably high levels of cooperation, while the resulting topology alone, used as a static substrate, does not support cooperation at all in the Prisoner's Dilemma.

8.7 Conclusions

In this chapter, we have studied another dynamical model for evolutionary game dynamics in a growing, network-structured population [16]. In contrast to most models for evolutionary games on dynamical networks that consider a constant population size [17–23], these networks grow. Nonetheless, individuals cannot break links and cannot control directly how many new individuals will establish connections with them. The two main changes we have made in this new model, with respect to the evolutionary preferential attachment studied in Chap. 7, are on the one hand, that now the probability of attachment is exponential with the payoff of the node, and on the other hand, that we allow irrational strategy changes, by using a Fermi-like function for the probability of changing the strategy.

One important difference that has been found is that under strong Payoff Preferential Attachment ($\alpha = 1$), the topology of the networks generated are dominated by the presence of a few super-hubs, which attract most of the links of the rest of the nodes. The existence of very few hubs and a large number of poorly connected nodes in network models have been widely reported before [24]. In fact, it has been shown that when networks are grown following a non-linear preferential attachment rule of the sort $p_j = \frac{k_j^\nu}{\sum_{l=1}^N k_l^\nu}$, with $\nu > 1$, star like structures are obtained [25]. Here, we have shown that the same kind of networks can be produced when the dynamics driving the attachment process is dominated by the most successful players.

Even when Payoff Preferential Attachment is not too strong (for instance, for $\alpha = 0.1$), super-hubs emerge, a clear mark that successful players are likely to attract many of the links of the new nodes. If newcomers preferentially attach to the successful players in the game, then high levels of cooperation are possible. But this cooperation depends on the growth of the network, the population structure alone will not lead to such high levels of cooperation. Thus, payoff preferential attachment differs from the usual promotion of cooperation in structured populations. In particular, it has been shown that heterogeneous static structures favor cooperative behavior due to the existence of hubs. However, as Fig. 8.6 shows, the presence of super-hubs is not enough to sustain cooperation in the networks grown following the scheme discussed here.

In other models, the probability to adopt a strategy that performs worse than your own is zero [26, 27] (see also some previous chapters). In particular, together with synchronous updating of strategies, this can lead to evolutionary deadlocks, i.e. situations in which both strategies stably coexist. Here, we have adopted an update scheme in which individuals sometimes adopt a strategy that performs worse. Due to the presence of such moves, sooner or later (often much later) one strategy will reach fixation. It is to say, the final state of the systems discussed here will be inevitably all-C or all-D. However, when β and the ratio b/c are large enough, both cooperation and defection can coexist for a very long time.

We also want to remark that our growth mechanism has another interesting feature: it has been shown that the average level of cooperation obtained in static, scale-free networks, is robust to a wide range of initial conditions (see Chap. 3). However, for the networks grown using the Payoff Preferential Attachment, the initial average number of cooperators in the neighborhood of the super-hubs determines the fate of cooperation in the whole network, leading to a much more sensitive dependence on the initial conditions of the system. This has been proved by the huge drop of cooperation in the system after some time steps, once we have reinitialized the strategies randomly among the individuals when the full size had been achieved. From this point of view, the weak dependence on the initial conditions reported in static scale-free networks is not trivial.

Finally, we point out that it would be of further interest to study the model discussed here with other 2×2 games. As we have shown, the game dynamics seems to have a weak impact on the structure of the resulting networks. Whether or not this holds in general will elucidate the question of the influence of different games on the network formation process.

In summary, the model studied in this chapter shows that the interplay between the game dynamics and the network growth leads to complex network structures. Moreover, not only the structure of the interaction network is important for the evolution of cooperation, but also the particular way this structure has been obtained. Our work shows that playing while growing can lead to radically different results with respect to the most studied cases in which game dynamics proceeds in static networks (which is in fact a conclusion we also made when studying the model of Chap. 7).

References

1. L. E. Blume, Games and Economic Behavior **5**, 387 (1993).
2. G. Szabó and C. Tőke, Phys. Rev. E **58**, 69 (1998).
3. C. Hauert and G. Szabó, Am. J. Phys. **73**, 405 (2005).
4. A. Traulsen, J. M. Pacheco, and M.A. Nowak, J. Theor. Biol. **246**, 522 (2007).
5. A. Traulsen, M. Nowak, and J. Pacheco, Phys. Rev. E **74**, 011909 (2006).
6. J. Pacheco, F. Pinheiro, and F. Santos, PLos Comput. Biol. **5**, e1000596 (2009).
7. F. C. Santos and J. M. Pacheco, J. Evol. Biol. **19**, 726 (2006).
8. A. Pusch, S. Weber, and M. Porto, Phys. Rev. E **77**, 036120 (2008).
9. S. N. Dorogovtsev, J. F. Mendes, and A. N. Samukhin, Phys. Rev. Lett. **85**, 4633 (2000).
10. A. Rapoport and A. M. Chammah, **Prisoner's Dilemma**. (Univ. of Michigan Press, Ann Arbor, 1965).
11. R. Axelrod, *The Evolution of Cooperation*. (Basic Books, New York, 1984).
12. M. Nowak, Science **314**, 1560 (2006).
13. M. A. Nowak and R. M. May, Nature **359**, 826 (1992).
14. T. Antal and I. Scheuring., Bull. Math. Biol. **68**, 1923 (2006).
15. H. Ohtsuki and M. A. Nowak, J. Theor. Biol. **243**, 86 (2006).
16. T. Gross and B. Blasius., J. R. Soc. Interface **5**, 259 (2008).
17. B. Skyrms and R. Pemantle., Proc. Natl. Acad. Sci. USA **97**, 9340 (2000).
18. V. M. Eguíluz, M. G. Zimmermann, C. J. Cela-Conde, and M. San Miguel, American Journal of Sociology *110*, 977 (2005).
19. M. G. Zimmermann, V. M. Eguiluz, and M. S. Miguel, Phys. Rev. E **69**, 065102(R) (2004).
20. J. M. Pacheco, A. Traulsen, and M. A. Nowak, Phys. Rev. Lett. **97**, 258103 (2006).
21. J. M. Pacheco, A. Traulsen, H. Ohtsuki, and M. A. Nowak., J. Theor. Biol. **250**, 723 (2008).

22. S. V. Segbroeck, F. C. Santos, T. Lenaerts, and J. M. Pacheco, Phys. Rev. Lett. **102**, 058105 (2009).
23. J. Gómez-Gardeñes, M. Campillo, L. M. Floría, and Y. Moreno, Phys. Rev. Lett. **98**, 108103 (2007).
24. F. Leyvraz and S. Redner., Phys. Rev. Lett. **88**, 068301 (2002).
25. P. L. Krapivsky, S. Redner, and F. Leyvraz., Phys. Rev. Lett. **85**, 4629 (2000).
26. F. C. Santos and J. M. Pacheco, Phys. Rev. Lett. **95**, 098104 (2005).
27. L. M. Floría, C. Gracia-Lázaro, J. Gómez-Gardeñes, and Y. Moreno, Phys. Rev. E **79**, 026106 (2009).

Chapter 9
Summary

Finally, as a conclusion of this Thesis, we will now summarize the main results we have obtained in the different parts of it. We have addressed the study of the evolution of cooperation on complex networks, using among the different social dilemmas, mainly the Prisoner's Dilemma game as a metaphor of the problem, analyzing the possible outcomes of the dynamics, depending on the underlying topology.

In the First Part of this work, we have focused on the evolutionary approach to several 2×2 games on static complex networks. In the first chapters we have studied the evolution of cooperation in ER vs. SF graphs when playing the weak Prisoner's Dilemma (Chap. 3), and the general Prisoner's Dilemma and the Hawks and Doves game (Chap. 4). We have confirmed the fact that cooperation benefits from degree heterogeneity, and we have also checked the differences between random and scale-free networks, as far as the microscopical organization of the strategies is concerned. Specifically, while for the first kind cooperators form several clusters, for the latest, they congregate in one single cluster that always includes the hubs. Moreover, we have studied the influence of the initial concentration of cooperators, ρ_0, on the final outcome of the dynamics, finding also important differences between both topologies: for random networks, there is a threshold in ρ_0 under which, cooperation does not survive, while for scale-free systems, there is always a non-zero level of cooperation, as long as $\rho_0 > 0$.

On the other hand, in Chap. 5 we have performed an study focused on the Prisoner's Dilemma game on random SF networks. These structures are obtained by rewiring conventional BA-SF graphs, in a way that destroys any node-node correlations. We confirm the fact that random SF networks can not sustain such high levels of cooperation as BA ones. Thus, the absence of age-correlations is a crucial factor in the dynamics of cooperation. Besides, we have found that the organization of cooperation into clusters is very different from the well-know case of BA structures. In this case, instead of forming only one cluster, cooperators form several of them, being in this way more vulnerable to the attack of defection. This result is a consequence of the fact that there are no age-correlations: the oldest nodes (that are usually the hubs) are no longer connected forming a single cooperation core, but they are

J. Poncela Casasnovas, *Evolutionary Games in Complex Topologies*, Springer Theses,
DOI: 10.1007/978-3-642-30117-9_9, © Springer-Verlag Berlin Heidelberg 2012

apart, forming several. Thus, this work can be consider a useful null-case for new studies to come on SF networks. Moreover, we have made an analytical approach to the problem, using a mean-field treatment, with a further compartmentalization of cooperators and defectors into degree classes. The results render by this calculations do not coincide with the numerical simulations performed on top of random SF networks in the general case. Nonetheless, we have considered a particular set of initial conditions, where we have assigned as cooperators all nodes with connectivity higher than a given value k^*. In this particular situation, we can say that our calculations qualitatively agree with the simulations. We consider that this work can be expanded to a more general scenarios.

In Chap. 6 we have approached the problem of cooperation in SF networks from a new perspective. We are not aware of any previous work addressing the study of cooperation in such topologies when restraining the number of interactions per node and per round of the game, but leaving the degree distribution of the system intact. Thus, we have considered this new situation, analyzing the level of cooperation achieved by the system when individuals are engaged on the Prisoner's Dilemma game but they are allowed to interact in every round of the game just with k^* of its topological neighbors. We have found some interesting results, such the fact that the highest level of cooperation for a given value of the ratio cost-benefit is not in general for the case of unrestricted interaction, but for the situation where some kind of constrain is imposed to the nodes. This is a somehow unexpected result, because it is well known that heterogeneity favors cooperation, but now we introduce a further consideration, restricting the number of interactions up to an extend renders even higher levels of cooperation in SF topologies.

On the other hand, we know that real networks are not static entities, but they evolve in time. New nodes and links may be added or removed, so the structure keeps changing, and this fact will most certainly affect the outcome of the processes developing on top of them. Thus, in the Second Part of this Thesis, we wanted to take into consideration that idea, studying systems where not only an evolutionary dynamics is taking place, but also the network itself is growing, and this growth is connected with the outcome of the dynamics. Specifically, we have developed two models where we combine the growth of the network and the outcome of the game in which the nodes are engaged. Although the basic idea is the same for both models, they differ in some particular details.

In Chap. 7 we have presented the first model, where growth and dynamics are entangled. In this case, the probability of an existing node to get a link from a newcomer node is a linear function of its fitness (understood as the payoff accumulated during the last round of the Prisoner's Dilemma game). We can tune the relative importance of the fitness in this process with a parameter that represents the intensity of selection. In this way, it can range from a random process, where the outcome of the game is irrelevant for the attachment of the nodes, to a strong selection situation, where it is the decisive factor. On the other hand, the updating rule chosen for the dynamics is Replicator-like, so the probability of a node for changing to its neighbor's strategy is proportional the difference between their payoffs. Using this model, we have analyzed some relevant topological properties of the systems, depending on the parameters. Thus, we have found that for the weak selection limit, we obtain random structures, and SF ones arise when imposing strong selection. Moreover, this model allows us to obtain several intermediate topologies, as far as the degree distribution is concerned. We have also found that some of the networks created with this model share other topological features with real networked systems, such as the power-law dependence of the clustering coefficient with the connectivity of the nodes. Regarding the evolution of cooperation, we have found that these systems can sustain very high levels (even higher that the BA structures). What is more, once the growth has ended, the system evolve to even higher levels of average cooperation. This fact is rooted in its particular microscopical organization of cooperation among connectivity classes: these systems display some important differences with respect to cooperation in complex static networks such the fact that now hubs can be stable defectors in the long term. To the best of our knowledge, this is the first time that stable defector hubs are reported.

Finally, in Chap. 8 we have exposed the second model, where there are two fundamental changes with respect to the first one: on the one hand, the probability of attachment is an exponential function of the fitness, and on the other hand, the updating rule we have used is Fermi-like, so now, we allow irrational changes of strategy (a node can imitate a neighbor that has obtained less benefits). As the the previous model, we have analyzed the resulting structures and the outcome of the dynamics. The topologies arising from this second model can be not only random or SF, but also star-like, due to the exponential dependence mentioned before. Nonetheless, regarding the evolution of cooperation, we have found that these systems do not promote cooperation as well as the first model does.

About the Author

Julia Poncela Casasnovas studied Physics at University of Zaragoza (Spain), where she received her Ph.D. in 2010 under the supervision of Dr. Yamir Moreno, Dr. Jesus Gomez-Gardenes and Dr. Mario Floria. She is interested in Statistical Physics, Non-Linear Physics and Complex Systems in general, and, more specifically, in different processes on top of complex topologies, such as social networks. She joined the Amaral Lab (Northwestern University, US) in November 2010, where she works mainly on modeling and analyzing social systems.

J. Poncela Casasnovas, *Evolutionary Games in Complex Topologies*, Springer Theses
DOI: 10.1007/978-3-642-30117-9, © Springer-Verlag Berlin Heidelberg 2012

CPSIA information can be obtained
at www.ICGtesting.com
Printed in the USA
LVHW011500280620
659213LV00003B/46

9 783642 301162